There and Back

Magnus Pyke is famous for his way of explaining science on the TV screen and over radio and as a result is much in demand as a broadcaster and lecturer all over the world. He can bring life and wit to the most difficult subjects.

He has a string of degrees and learned qualifications, which he wears lightly; twenty-five years' experience as director of a research establishment; four years as Secretary of the British Association, initiating working parties on subjects of vital public concern and collating and presenting the facts of worldwide scientific advance. He is also one of the few distinguished scientists who has inside experience of Dartmoor Prison – in diet research.

He has an inexhaustible appetite for the curiosities of man's scientific progress and an unshakable belief in a truly scientific approach to life as the only safe way forward.

Also by Magnus Pyke
in Pan Books
Butter Side Up!

Magnus Pyke

There and Back

illustrations by Paul Sharp

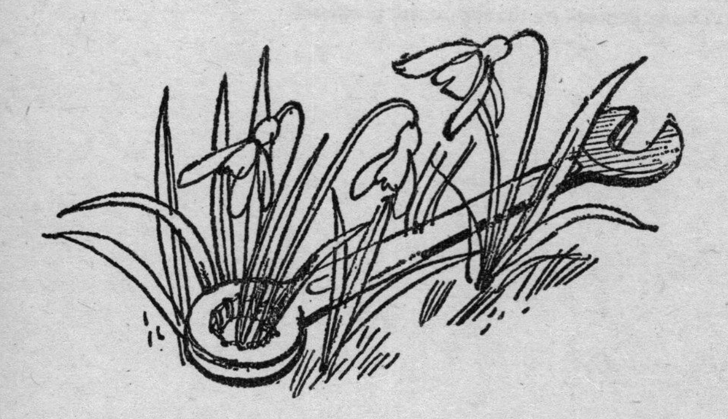

Pan Books London and Sydney

First published 1978 by John Murray (Publishers) Ltd
This edition published 1980 by Pan Books Ltd,
Cavaye Place, London SW10 9PG

ISBN 0 330 26020 2
Printed and bound in Great Britain by
Richard Clay (The Chaucer Press) Ltd, Bungay, Suffolk

to

Nicolas-Léonard-Sadi Carnot

from whose work came
fish-fingers, air-conditioning
and artificial satellites

Acknowledgements

My thanks to the Executors of A. P. Herbert for permission to quote the verses on p.162, to Paul Sharp for gracefully illustrating my thoughts, and to Phyllis McDougall for indexing them.

M. P.

Contents

1 Coming and going

How happy we are to be able to live a comfortable, warm, wealthy life in the modern age of today. Grateful to science we may not be but, if we ever stopped to think – which we seldom do – we'd know that we ought to be. Spinsters no longer have to spend hour after hour spinning to make the thread to weave the sheets we sleep between. No one has to labour any more in boggy fields to grow the flax for the linen for those few rich people who can afford linen sheets. As it happens, making linen is a disagreeable smelly business: the flax has to be cut down and left to rot in the fields so that the fibres can be beaten out of it, a process called wretting. It was probably the second most disgusting process of the pre-scientific age, exceeded in disagreeableness only by the tanning of leather where hides had to be soaked for weeks in a slush of dog manure, strangely enough the richest source of the particular enzyme that most effectively softens skins. Those whose fortune it once was to live downwind of a tannery can confirm how potent was its effect on what used to be called the atmosphere but is now known as the environment.

And if the once prized linen sheets were superseded by cotton ones, cheaply available for our lucky ancestors from the labours of slaves in cotton plantations, even these imposed a burden on washerwomen – and they needed ironing as well. Beautiful smooth nylon sheets need none of this. And if we still sleep in cotton sheets, flat-irons no longer have to be heated on the fire, one left on the hot coals while the other – cooling all the time – was in use, to be slipped out of its metal shoe and exchanged every few minutes. Today, in these clever times, an iron naturally means nothing else than an electric one.

And just as electricity is piped in as a matter of course to every house, so is sewage pumped out. Back in those barbarous days of Charles II when Isaac Newton, the greatest scientist England – or perhaps the world – ever produced, first grasped what kind of universe it is we live in, poor unfortunate Pepys had his whole basement flooded with ordure from the overflow of his next-door neighbour's cesspool. What a wonderful piece of work is a sewage disposal plant! Yet how often do we, the inheritors of so much scientific skill and effort, exclaim in the fullness of our hearts as we press the handle and set the whole mechanism in motion? 'Go to the ant,' said the Victorian moralist to his pupil: what better phrase could the modern teacher employ than 'Go to the sewage farm!' There shall you find not only scientific enlightenment, not only a way to express gratitude for all the great benefits of science, but also moral enlightenment and a way to live in peace among the complex stresses of a crowded world.

To the sewage works come the undigested fragments of the fruits of the earth, crumbs – it could be said – from the rich man's table. Mingled with them are the bodies of dead creatures, the micro-organisms which we all carry around in our guts, to live with us and serve us, make us vitamins and chew up some of what we cannot effectively chew. Then, when we are done with them, we throw them out to die. Down the pipe they go to where on a specially constructed waterfall made of plastic honeycomb – it used to be a circular castle of coke – the juices of sewage serve as a medium for a complex United Nations of bacteria. One nation consumes one component of the brew and another another, and so on. These nations live together in harmony, if the sewage-works operator knows his science. But not only are there nations of bacteria, there are large creatures as well. Things that wriggle and paddle themselves along (under the microscope, of course) with rows of 'oars'. There are larvae, too, which in due time turn into insects and fly away.

I once had a colleague, Bob McGaw, a microbiologist, who was researching into the biology of effluent-disposal plants (sewage works to you). Peering down his microscope, he observed the little bacteria, and the not-so-little protozoa rowing themselves busily about. All at once, his eye peering down through the eyepiece saw looking up at him – another eye. Within the single span of a trickling filter-bed in a sewage works we have a dramatic model of what one day might be a harmonious world where the lion and the lamb would not only be lying down together but would be working together and, best of all, working together for the benefit of man – a living parable of the way man might usefully guide his own behaviour. There, all together on Bob McGaw's microscope stage, were bacteria and protozoa, larvae and flying insects all single-mindedly, while going about their own business, serving to use, break up and hence dispose of the wastes of urban man. It was almost as if the creature looking up from below was nodding its head at the man looking down from above to congratulate him on having, through his scientific understanding, so arranged affairs that London of the twentieth century should be more civilized than the London of the seventeenth.

A sewage works is such *fun* quite apart from its being so fine a monument to science. The sieves and strainers by which the larger odds and ends are sorted out are ingenious if not particularly educated, but after the first set of united nations I have described there is another world. Further compounds of different sorts unsuited to the air-breathing citizens of the trickling filter are passed into a closed chamber in which no air is allowed. Here they are set upon by a new combination of creatures, living their life in a different way, who obtain energy for life from the ingredients remaining in the thin consommé of sewage and leaving behind, not the carbon dioxide of us air-breathing animals, but a hot breath of hydrogen and marsh gas. Naturally, the ingenious operators of the sewage farm – applied scientists to a man – make use of these gases to keep the wheels turning and operate gas engines to run the pumps, heat the buildings and cook the meals in the sewage-works canteen.

Every aspect of today's delightful life bears witness to the debt of gratitude we owe to science. Coming down to breakfast, no longer do we need first of all to clean out the grate, lay the fire with dry sticks and set the kettle on to boil. As a child, I remember just this being required. The wonder of those fat towering structures we all (incorrectly) called gasometers, by which every town of any pretensions to up-to-dateness was dominated, was yet to come. Today, the giant Gothic – if slightly rusting – erections have lived their brief lives and will disappear as further advances in applied science increasingly allow gas to be distributed at high pressure. And who is there so lacking in imagination as not to feel a sense of wonder each time he turns the tap? To think that the secret, potent, invisible vapour by which the bacon is being fried has come from the deserts of Algeria on the borders of the Sahara or from the depths of the sea! Who would have conceived that the dry scholars in knickerbockers – figures of fun, scrambling about the countryside with little hammers collecting samples of rock – would so have developed the dilettante science of geology that by a bold amalgam of thinking and adventurous exploration, the subterranean stores of fuel would be exploited for our daily use?

'The ploughman homeward plods his weary way,' wrote

Thomas Gray, and if it was raining, the ploughman got home soaking wet. In those days of the eighteenth century macintoshes had not been invented and science had not achieved the wonders of plastic raincoats by which outdoor existence in a wet climate has been made so vastly more comfortable. Jane Bennett set out from her home to have dinner with Miss Bingley and Mrs Hurst barely three miles away; since her father could not spare the carriage, she was compelled to go on horseback. It rained, she became wet through and contracted a chill which kept her in bed for a fortnight. What if at the time Jane Austen wrote *Pride and Prejudice* Miss Bennett had had a snug, dependable automobile to take her to the party or at least a plastic mac, rubber boots and a modern umbrella?

It is said of Caesar Augustus that when he arrived in Rome it was a city of brick and when he left it was a city of marble. Modern citizens everywhere in the industrial world live in cities of marble – or at least, if not of marble, in the next best thing – concrete: hard, white, durable, strong, far easier and quicker to erect than stone, infinitely variable and far more versatile than any stone available to Augustus. Not only can we construct our buildings from this material – the product of chemical knowledge and understanding, of crystallography and rare insight into molecular structure – but the roads of the day, the great motorways linking together all the cities of the land can be made from it as well. The Romans with infinite labour constructed some few military arteries across the countryside, but people the world over had to put up with the mud of winter and the rutted dust of summer. Even in the favoured twentieth century, the majority of mankind, to whom poverty, remoteness or war has denied the blessings of science – for example in large areas of South America, Asia and Africa – are still plagued by the arduous labours of travel; having neither the knowledge of how to prepare concrete, nor the raw materials, the equipment and fuel needed in its manufacture, nor the great earth-moving and mixing machines without which road-building is of almost all human endeavour the most laborious. For us, the happy few, it's into the car and away.

We who through the blessings of science enjoy the privileges of mobility show by our actions how deeply we value it. Able to leave the towns we live far away from the factories and counting-houses where we earn our living. There is no compulsion on us to do this. As individuals we opt to live away; as communities we listen to the local authorities whose duty it is to be our corporate voice when they decree that factories *shall* be at a distance from dwelling houses. We take advantage of the gift of science and the technology springing from it, to *commute*, as we quaintly put it, miles away from home each morning and back again to the place we started from every evening.

Through the ceaseless drive of research the petrol engine becomes more powerful and sophisticated each year and the thirst to enjoy its blessings is most dramatically exemplified in our pleasures. Before the age of science and technology ordinary working people had to stay at home. On feast days they could enjoy the fair on the common, take part in the great spectacles of the church such as the miracle plays illustrating the stories of the Bible, go to the local theatre, or indulge in the conversation of the alehouse. If he was rich, a young man could hope once in his life to indulge a pleasure in foreign travel if his father was able to send him and his tutor on the Grand Tour to complete his education. As the benefits of science grew and became more widely diffused among all the members of the community, there came the day trip to Margate or, for Glaswegians, the steamboat trip out into the estuary of the River Clyde, 'Doon the Watter'. This extended to the week away from home at the seaside and now, in an age when the intellectual achievement of flying has become a commonplace, no member of the community is denied the enjoyment of travelling – 300 at a time at 600 miles an hour – from their homeland to a foreign resort, there to be housed in system-built dwellings, fed, amused and returned home with equal economy. Nor does the thousand-mile journey, which would have involved our ancestors in months of planning, trouble, some danger and substantial expenditure, cause them any effort at all. Through the blessings of applied science, they need only wish and the journey is accomplished with little more activity on the part of the traveller than if he or she had been an inanimate parcel. Indeed, so complete a mastery over distance has science provided that people who are old and poor can be supported in their poverty just as conveniently and economically in an out-of-season hotel in another country as in their own houses. The plane journey there and back is a bonus the richness of the times can easily supply.

Whether or not an urge towards locomotion is one of the built-in physiological desires of man as a species, it is clear that once the freedom to travel about at speed becomes available to a modern community, the members of such a community immediately exercise, as almost their most highly prized desire, this freedom. A number of philosophers have attempted to identify what seems to them to be the basic human motives: life, liberty and the pursuit of happiness have served in America for two hundred years and can continue to serve still. Even if these vague terms do not explicitly comprise what millions of people show by their actions *they* would expect them to include, namely the right to drive up and down in a motorcar. All modern states, in which science is well developed as a branch of public knowledge and is applied in the manufacturing industries of the nation, so arrange their affairs as to produce as many motorcars as they can for the enjoyment of the citizenry. Loosely administered countries such as the United States pride themselves on the fecundity of their automobile manufacture; rigidly ruled nations such as Nazi Germany held the production of huge numbers of Volkswagen – the people's car – as their highest achievement; newly industrialized nations such as Japan measure their advance towards success and happiness in terms of the increasing numbers of automobiles they succeed in making; countries in economic distress such as Great Britain contemplate with some measure of resignation the decay of their basic industries and the dismissal of their workers, but they do not abandon their capacity to produce automobiles. Veritably, the truth of the aphorism 'What is good for General Motors is good for the USA' transcends national boundaries and changing political attitudes. And of all articles in popular use in the modern age, the motorcar illustrates most vividly the direct relationship between the advancement of science and the increase in human happiness. Each year as the new models come out, the lights are brighter, the engine more efficient, the upholstery more durable and beautiful, the paintwork more delightful, the rear windscreen more subtly heated. And each of these is an added joy to the car owner.

Then there is the telephone, that splendid engine of human togetherness. How can one avoid becoming lyrical at the thought of what it makes possible. Should a man's car break down on the motorway, there at convenient intervals are telephones through which his call for assistance can be heard by those waiting to help him. How, we may well ask, could our ancestors have endured existence when their only means of asking a man from the next town to dinner was to send a message to him by a servant on a horse? By what miraculous means could the young men of an earlier age indulge – as literature tells us they did – their taste for women, deprived as they were by the backwardness of the times of any possibility of asking each new-met friend for her telephone number? How could business be transacted, bets placed, taxis called, burglaries reported to the police without telephones?

We have lived in an age of giants undreamt of by our predecessors. We are the first generation of mankind to have succeeded in stretching by our scientific capability the art of gunnery to the point when a projectile has been fired up so high into the sky that it has never come back. Above our heads it swings round the earth as it turns, like a conker on a string swung by a boy. We have put up several of these metal projectiles so delicately aimed that one hovers for ever over the Atlantic, another over the Pacific. You historians who write of the achievements of this splendid generation – how can we fail to delight in our godlike powers? – we who as a scientific community put up the first satellites, tiny moons all our own. Nor are they sterile and crumbling like the old worn-out moon God put in the sky before we were born.

Each of ours carries a telephone exchange so that we can dial straight through from Tottenham to Tokyo and find out what is happening on the Japanese stock exchange.

As we follow modern man through the delights of his day, the prizes won by modern science and the practical processes built on to it – few more than a hundred years old and many of them less than fifty – are apparent on every side. The chemical industry of the nineteenth century made nations rich, to be sure. But the great successes, such as the ways of making bleaching powder or sulphuric acid, did not much affect the quality of life, unless it was to make those parts of the town where the labouring classes lived smelly. The hearts of ordinary people do not beat faster nor are their cheeks flushed and their step more buoyant at the thought that a new and improved process for the production of sulphuric acid has been developed. One has to be an economist, a businessman or a chemist to know what sulphuric acid is needed for. There is a general misty understanding that, somehow or other, the prosperity and wealth of a nation can be measured in terms of the tons of the stuff it turns out, but few people know that it is an indication of the size of a nation's metallurgical industries. Bleaching powder was the key which unlocked the riches of India to the British Empire by making it possible for the mill owners of Liverpool to flood the Indian market with cotton textiles, but this achievement is a long way removed from the fruits of science which our modern citizen enjoys as he walks dry-shod on shoes with tough plastic soles, made of a substance derived from chemically polymerized petroleum.

We sit down on 'leatherette' seats (also an outcome of the chemical industry) on our swivelling office chairs supported on legs plated with untarnishable chromium, a wonder unknown to our ancestors. Our pleasure is enhanced by the knowledge that no longer do workmen need to rise at five o'clock to spend long hours at low pay making the wooden office stools and roll-top desks of long ago. Today, they too enjoy mopeds, cars, gas stoves, electric light, and if not telephones for all at least television for all; and they are able to travel and take holidays as long as their masters'.

Not all citizens need to work even in the comfort of their legally protected, warm, well-lighted offices and factories. So great is the wealth technological ingenuity produces that the community can by law decree that no young person need start life's work until he is past the age of puberty – or even later, because puberty comes early these days. Up till then the young can enjoy a life of instructional play: they are carefully isolated from any contact with workaday reality. Former generations of children learned at an early age that what they enjoyed was only what their fathers could afford to provide, and that one day on them too would fall the duty of providing support for *their* children. Even during the middle three-fifths of their life span, the citizens of an up-to-date community are only expected to devote themselves to productive endeavour for 231 out of the 365 days of the year (that is approximately five-eighths). Compared with the length of time which a man would devote each day to work if he were determined to do his utmost in a hard and hostile world, the science-supported man of the present works for two-thirds or less. Truly he is fortunate in needing to devote to work barely a quarter of what would be a full working year. Before he starts he has nearly 20 years of undemanding youth; after he finishes there comes a decade or more of active serenity during which politicians of every colour wish him well in their every speech. He has been freed from slavery.

It is a strange truism that in spite of the wealth, comforts, protection, benefits and amusements which science-based technology so palpably provides, people are not grateful, nor are they proud of the achievements of their scientists and engineers. In spite of all the things they can use, regardless of the fact that people can live and die with much reduced pain, and regardless of the fact that nobody dies from smallpox, cholera or plague any more, people are not happy. I have hinted at a paradise attained, but the picture is not a true one; not because modern citizens do not have at their disposal all the technological facilities I have described and more, but because they do not appreciate and enjoy them – they take them for granted, to be sure, but little more.

'What a miserable sinner I am,' our grandparents used to say. And they *were* miserable too: the more virtuous they became and the less important their sins actually were the more miserable they felt they ought to be. There is no exact evidence to prove the point, but it is a reasonable assumption nevertheless that the really serious sinners of the time, the perpetrators of murder, arson, rape and pillage, were less miserable and more cheerful than Victorian fathers of families whose sins were confined to dropping a foreign coin into the offertory plate at church or instructing the parlour maid to tell a tedious visitor they were out when they were actually in. And the same principle appears to apply to today's populations where their blessings are not of grace but of science.

The first conversion – from a draughty, slow, unreliable and expensive horse carriage to a reliable (or, at least, comparatively reliable), swift, warm, comfortable and economical motorcar – brought with it joy and delight. And each successive advance, enabling the vehicle to go further; to continue going longer between punctures; and, most exciting of all, to travel faster and faster; each was initially treated with the applause it deserved. Soon, however, the delight became modified. The increased capacity for speed was found not to bring unalloyed satisfaction. A relationship was noticed between an increase in the speed of travel and the number of people killed in road accidents. The graph may not be a simple straight line in which twice the velocity is accompanied by twice the number of casualties. For example, a regulation restricting the maximum speed of heavy lorries to 20 miles an hour, instead of reducing the number of accidents on the road may actually increase them: frustrated drivers of faster vehicles may be found to involve themselves and other people on the road in greatly increased risks in their endeavours to get past the crawling lorries. Nevertheless, the graph of death in relation to speed is a positive mathematical relationship. At one end, when the speed is reduced to zero and nobody moves at all, nobody is killed by motorcars. At the other extreme, when the Formula 1 racing driver pushes the speed beyond what the car can manage, the machine flies to pieces and the driver is killed. Between these two lies a steady increase in risk. Consecutive advances in technology – first 4-wheel brakes, then discs; improved tyres; discoveries in metallurgy, rubber technology and hydraulics leading to better springing; more reliable bearings and more controllable engines – each one of these produced a change in the detailed topography of the graph, but its general trend remained unaffected.

Plotting a curve relating speed, which is an achievement of scientific and engineering progress, against a merely human objection to being killed, which is nothing more than a statistical index of systems efficiency, leads to the intrusion of a non-scientific factor into the equation, be it the wish to stay alive or a taste for roast beef. This subjective influence for the first time induces a change in the basically unanswerable proposition that the step-by-step application of science to public affairs makes things progressively better and better. Where formerly it was accepted that to travel at 20 mph in an automobile is better than to travel at 8 mph on a horse, and that to travel at 60 mph in a better motorcar is better than to doddle along at 20 mph in an old-fashioned one, people have changed their ideas and decided that to travel at 100 mph, all things considered, may not be better than to go at 70. In fact, they are opting *not* to accept the further technological possibilities which, in this field of transport, the advancement of science can offer.

The picture that I sketched a little while ago of the joy and convenience offered by the telephone is in its main essentials true. But as the possibilities of communication become more and more extensive each dwelling house possesses not merely a single telephone – once to be found inconveniently fixed to the wall in a draughty hall – but a telephone in every room, including the bedroom; and then phones extend their territory beyond the house and out into the garden; and then more and more cars are fitted with them. They begin to appear with the disposable paper napkins at table in some restaurants; in the more advanced cultures their sensors become an integral part of people's clothing, as a 'bleeper' to call the wearer to an instrument elsewhere; but, almost as we watch, the bleeper becomes transformed into a miniaturized radio-telephone such as the police use. And finally, as all this happens, the citizen begins to feel that he can have too much of a good thing: the telephone has been converted from a boon into a nuisance; instead of making life better, its very ubiquity and effectiveness make life worse. The marvel of telephony advances swiftly from the electric current pulsing out the dots and dashes of the Morse code to the dial telephone with the intellectual brilliance of its computerized automatic switching; it is supplemented by radio waves shot from the lofty convoluted minarets which dominate each city and which call the faithful, not as they did once to prayer to think about good and evil and the purpose of life, but to business appointments by which they can acquire money or bet on the speed and agility of a horse which they have never seen – and it becomes apparent that in telephony too a standstill or even a retrogression might be preferable to the advancement of science. How many people there are who, if posed with the question of what they would most dearly love to escape if they were removed from the pressures of modern life and sent to live on a desert island, express their pleasure at the idea of being relieved from the insistent demands of the telephone? A point can be reached when electronic communications can become too good.

There are a number of other pieces of evidence to show that although the first application of science to the affairs of daily life adds to the sum total of human happiness, there appears to be a point when, after a plateau in happiness has been reached, further additions of science and technology, far from bringing the further blessings they were designed to produce, cause the total of happiness to fall.

There is another way in which too many gifts of science can fall into the lap of a modern citizen for his own good. This can come from the inexorable need for a community – any community, rich or poor, urban or rural – to pay the social cost of what they get. The capability – matchless in any other period of history – of being able to fly across the world to share the love of distant relatives, to do business, capture fleeing criminals, collect news, visit distant lands and see strange sights, must be paid for in terms of the conversion of fields of corn and meadows rich with daisies (where birds sing and butterflies flutter on painted wings) into ugly prison-exercise-yard stretches of infertile tarmac surrounded by the noisy uneasy restlessness of terminal buildings and hangars; the whole being the central pustule of a wider area of countryside where the noise of aircraft renders the natural exchange of human conversation impossible. In spite of all that architects can do and that money can buy, in spite of the devoted efforts of those who work there to diffuse friendship, loving-kindness and civility no one has succeeded in making a modern airport into a place of voluntary, far less agreeable, social resort. On the contrary, its success is measured in terms of the speed with which people can get away from it. Yet the small grass airfield of early flying days, complete with daisies and buttercups, was a delightful place to linger beside.

A large airport, with all the inevitable inconveniences associated with it, is part of the price that has to be paid for the advantages of flying fast, one of the most brilliant technological achievements of man. This is the first period in the history of humanity when the benefits of rapid, effortless and substantially safe transport have been available. Ever since flying became publicly possible – which is little more than within the last half century, people in every land have eagerly welcomed each successive advance in science by which it has become safer, quicker, more comfortable and more efficiently organized. To cross the Atlantic in a sailing ship was a long-drawn-out and hazardous adventure. Men and women who committed themselves to the deep knew that they were subject to shipwreck, storm and tempest. The age of steam when it came was a giant step forward. But the steamship was an evanescent step in technological progress which the aeroplane quickly brought to an end. And during the short history of flying, aeroplanes themselves have improved beyond recognition until today the great jumbo jet bears little relationship to a bird. Instead it has become a crowded hotel lobby in motion.

The question now is whether the brains, skill and knowledge which have brought this public wonder into being are going to continue to make travel faster and more wonderful still. Or is the progress of science about to advance in a new direction? Shall we, instead of using more science, use less?

As soon as World War II was over in 1945, more and more people wished to fly in more and more, bigger and bigger aeroplanes from London to every part of the world. Science was brought to bear to provide radio beacons brilliantly applying the fruits of scholarship and discovery to guide the aircraft in and out by night and day, in calm and storm, sunlight, moonlight or fog. Soon, one after the other, the planes came in and out of Heathrow Airport until no more could be squeezed into their unbroken ranks. Another airport was commissioned, Gatwick, from which a second ladder could be built into the heavens up and down which another file of machines could climb. And when this too filled, the nation took thought to build a third. But the plateau, it seemed, had been reached. There was a limit to the urge to fly, there was a price – in noise, smell, acres of tarmac and cubic feet of central-heated waiting room where the voice of the tannoy carries ceaselessly on the stagnant air – which people were not prepared to pay. The people revolted and the third London airport was not built.

Nor is the appetite for the fruits of science saturated merely by the hardware and furniture of life that it can provide. The noblest gift of science is insight. Insight into the mechanism of the living flesh allows those who possess it to control life and death, insight into the structure of matter provides mechanical power. Most important of all, an understanding of science allows a community to foresee and make provision for the future. Once the census is taken, the demographer can forecast the future. The biochemist supported by an efficient pharmaceutical industry can control the fertility of the nation's women to adjust the future members of the community to the planned provision of dwelling units. Foresight, based on scientifically adequate data, can ensure a satisfactory future for all – a future of well-paved, well-lighted streets, adequate drainage, recreational facilities, teaching machines in the schools, shopping precincts and medical services. Well done, indeed. Just so was it done in the 1960s. Scholars sat down to think, computers computed, intellectual insight was based on scientific scholarship – truly the fruit of the Tree of the exact Knowledge of Good and Evil. Useful thought, godlike thought, the prelude to action, was taken to plan for the needs of the people who, it was confidently predicted, would inhabit the south-east corner of England fifty years ahead. It was calculated that all the area of land eastward of a line from the Wash to Southampton would have to be paved and built over with houses, schools, factories, hospitals and football stadiums together with their necessary services.

Again, it seemed, the price was too high. No one found fault with the plan. The data upon which the forecasts were based were impeccable, the logistics unquestionable but the original Adam, the Lumpen Proletariat – the people – moved by their feelings declined to accept the rich sanitary life they were offered.

Some time before, they had experienced on a smaller scale what informed technology could do to improve their conditions. In 1900, no one calculated how many cubic feet of air are needed for healthy sleep, how many square feet of window are required for the prevention of subclinical claustrophobia, what the diameter of the vent pipe of each domestic water closet should measure, because there were no water closets in many of the houses to measure. Streets were put up higgledy-piggledy and the people who lived in them added rabbit hutches, pigeon lofts and bicycle sheds. Scientific thinking brought to bear on how best to remedy the palpable deficiencies of an area built-up in this way led to the admirable conception of tower blocks, and system-built dwelling units. The specifications of the new apartments were better in every way than the old. And clearly, to replace the slums of fifty years with the planned design of an advanced technological society, the former must first be cleared away.

Radical development of a town, in which the whole area of existing buildings, houses, shops, pubs, theatres and churches is scraped clear leaving only the bare earth, can be demonstrated to be economical, efficient, sensible and, above all, scientific. The plastic drain-pipes, the cables, the water, the district heating can then all be set in their places, the recreational areas laid out and the new living structures put up. Thus, indeed, it was done in old tumbledown towns in the 1950s and the 1960s. Unfortunately, when the people came back to reinhabit the places where they were born they did not recognize them and, worse still – in spite of the heating, ventilating, sound-proofing, upgrading, all done on a proper scientific basis – they did not much like what they found. Thus it happened that later on the south-east corner of England never did get paved.

We can never have too much science. That is to say, regardless of how much is discovered about the universe and the stars in it, the stars we see with our eyes, our radio receivers or our x-ray sensors, there will always be more to discover. As many Nobel prizewinners as you like may discover DNA, the so-called double helix which is the chemical molecule of heredity; and, having discovered it, they may crack the 'genetic code' as much as they please. Even so, they will never know enough to construct a good man out of his component parts. Come to that, psychologists of the highest and most prolonged educational status will never be able to produce an exact specification of a 'good man'. So there will always be more science to do and more things to discover. What Newton said three hundred years ago is just as true today. He, the greatest of scientists, put it thus: 'I do not know what I may appear to the world, but to myself I seem to have been only a boy (he was over 80 at the time) playing on the sea shore, and diverting myself in now and then finding a smoother pebble or a prettier shell than ordinary, whilst the great ocean of truth lay all undiscovered before me.' Science will always go on as long as there are people possessed with divine curiosity and the urge to find out the truth. But the degree to which the ability such knowledge provides actually gets things done and affects practical affairs is a different matter altogether.

Few people want to die, and up till now it has been taken as axiomatic that the scientific knowledge by which killing diseases and complaints can be controlled should be made use of. But the axiom becomes less obvious when the causes of death are cirrhosis of the liver or cancer of the lung and the scientific treatments mean no drinking or no smoking. And the problem would be at its starkest if the scientific discovery were made by which everybody's life-span could be extended to 100 if they gave up eating and took their meals by means of a syringe. Even if the treatment were later modified so as to involve no inconvenience or changes in habits, there might be those who would rather not have to live until they were 100 years old.

This book is only partly about science. It is concerned only with such science as is needed for the basis of technology. It is only because science makes technology so potent that the last century or so has been radically different from anything that was ever seen before. The problem is: how much of this science-centred technology do we want? Or, to put the same question in another way, how much of this kind of technology is needed to provide a good life? With too little, people will feel themselves deprived of wrist-watches and transistor radios. With too much, they will start talking about the evils of civilization. They will suffer so severely from what were once called smell, dirt and noise that they will give them the most evil word they know – pollution. The mass of public hardware and the weight of private possessions will drive them mad, although they will have become so frightened of the stain of madness that they will call it 'mental disturbance'. Too much technology, too automated and mechanical a life full of passenger transit systems geared to transport people like cans of beans on automatic conveyers to their systematized work, their mass-produced pleasures, and thence to the mechanized dwelling units once called homes can, it seems, cause a relaxation in the bonds by which a corporate society is held together.

Too much of the rational spread of technology undoubtedly affects the way a community views the nature of life and the quality of behaviour. Bastards, like madmen, disappear in their once-familiar (if not always popular) guise only to reappear as the children of that biological anomaly, the 'one-parent family'; although it is not entirely clear whether they are much more welcome as such than they were before. In pre-scientific societies, deprived though they may have been of wristwatches and launderettes, people nevertheless did possess some idea of good and evil, sin and virtue, to guide their behaviour. I make no judgement as to whether the sum of human happiness was greater or less for this reason when the net injection of science was low. But in the super-rational world of high technology where the amount of applied science may have become too much, there is no more sin: the devil, more hardly used even than poor old Father Christmas, has vanished not even to survive as a tawdry advertisement for December shopping. In place of evil there is a nebulous entity, to be changed by every whim of fashion, every speech from parliament; what we call *anti-social behaviour*, under whose banner any unpopular action from keeping a shop ('capitalism'), to smoking, beating young children to death ('the battered baby syndrome, not a crime, an illness like impetigo') to sodomy or robbing a bank ('due to an unhappy infancy') are equal.

There is a balance sheet of credit and debit in applied science. For most of the last hundred years the evident credit balance has been axiomatic: the Industrial Revolution followed by the Scientific Revolution (of plastics, antibiotics and nuclear energy) would never have happened if people had not desired the power, wealth and comfort that modern science and technology provide. All at once, however, we have begun to realize that the balance sheet has changed. Do the gains of science still far outweigh any disadvantages such as dirt, smell, noise and ugliness? Although technology is quite capable of dealing with such practical problems as the corrosion of old buildings with sulphur dioxide in the atmosphere, or damage to the health of workmen by asbestos dust, it cannot attack a more deep-seated problem. The balance sheet with which we are really concerned is one which has plagued the rich since the beginning of history.

The poor are too busy trying to earn their living and to ameliorate their poverty to be bothered. It is the rich who are in trouble. It is this business of the good life, whatever that may be, that is the real problem. In the main we have solved the practical problems. When the politicians and the newspapers of industrialized countries grumble most about full employment and the devaluation of the currency they are less concerned with how to produce motorcars, processed food or television sets than how to arrange our affairs so that we can enjoy all these goods. Worse still, there is the subtle question whether we really do want what we think we want. Perhaps real happiness may not be derived from spending the summer days in noisy factories or soulless offices, having achieved the 'full employment' that everyone desires, in order to earn more and more money to buy more and more things.

We are the rich, beginning to realize that we can be surfeited by our riches. Are we going to put science in its place and set out in search of a more balanced society?

2 Prizes, two a penny

One of the comparatively few occasions from my schooldays that I still remember at all clearly was an argument with one of my seniors – a fellow pupil, as I recall – when he turned to me with devastating contempt to say 'And you call yourself a sportsman!' Even as he said it and I started to crumple, it suddenly dawned on me that I wasn't the kind of person he would call a sportsman, nor did I want to be. And at that very moment my education clicked forward one notch. This incident, rare, surprising, even refreshing, comes to mind now as I start to look back at a time when it was obvious that we were all sportsmen, all competing for prizes which everybody was striving to win. Some people, even if only a few, have suddenly awoken to the fact that they don't really want the prizes which scientific striving can bring. This represents a radical change in outlook. How remarkable such a change – if it came to affect the community as a whole – would be is best shown by looking back at some of the great prizewinners of the past. Just consider the strange careers of a Frenchman, Nicolas Carnot; a Scotsman, James Watt; and a German, Rudolf Diesel.

Carnot, poor chap, was only identified as a prizewinner after he was dead, but once his celebrity was recognized it was great indeed. Not only has he achieved immortality as the originator of what generations of bright A-level schoolchildren and harassed students have had to cram into their heads as the 'Carnot cycle', one of the basic principles of thermodynamics, but from this brilliant intellectual conception great things have emerged. Ice rinks, domestic refrigerators, frozen fish-fingers and, above all, liquid oxygen to fill the fuel tanks of rockets to the moon and intercontinental ballistic missiles alike depend on the knowledge Carnot won from nature.

The prizes Carnot won were of two sorts. First, those who are interested in pure thought would value his achievement in theoretical thermodynamics today as much as it was valued and praised more than a century ago. Second, there were the tangible prizes – the fish-fingers and the liquefied gases – but of these he knew nothing, so he would not have been able to guess whether or not we, his successors, would become tired of them.

To understand Nicolas Carnot's romantic tragedy one must go back to his father, Lazare Nicolas Marguerite Carnot. He was a professional soldier in the French army and before the Revolution he had, as a young officer, written several books on the art of fortification. Having taken up an unorthodox line on the controversy of what almost amounted to the 'Maginot' philosophy of static strong points contrasted to a 'Blitzkrieg' idea of rapid movement and attack he became so deeply embroiled with his superiors that he was actually imprisoned for a while. His time came with the Revolution and its immediate aftermath, when he quickly became the general responsible for military affairs on France's north-east corner. Here he achieved a number of remarkable successes partly owing to his strategy of rapid, controlled movement of his troops but partly because he was an excellent administrator and took great pains to ensure – as was not done by other military commanders of the day – that his soldiers were kept supplied with three essentials to efficiency: food, clothing and ammunition. By

1793 he was back in Paris reorganizing the whole of the Republic's military forces. He succeeded in avoiding the main political ructions of the times and by 1800 had become minister of war. With great efficiency he tightened up the administration of the French army and made it into the flexible instrument which Napoleon subsequently showed it to be: contrary to long-established custom, he refused to accept presents from army contractors. He was never much enamoured of Napoleon's campaigns of annexation and, as the good republican he had become, he did not particularly fancy his imperialistic ambitions. But during the Hundred Days when the Emperor had escaped from exile in Elba, he accepted responsibility for the defence of Brussels, which he carried out brilliantly. It is interesting to note that during the course of this turbulent life of high military and administrative responsibility he wrote several books on geometry, dynamics and other branches of mathematics. He died in 1823 at the age of 70. Nicolas, about whose work I want to write, was his eldest son.

He was a bright young man and, as was perhaps natural in view of his father's profession, obtained a commission in the engineering corps of the army after having obtained his professional qualifications at the Ecole Polytechnique in Paris. All seemed set fair for a glittering military career. Then came Waterloo and the end of the Republic. His father was disgraced and he himself, although not discharged from the services, was shunted into dead-end duties. With nothing much to do in his career and little prospect of real advancement he took up all sorts of other activities. Some of these were physical: he acquired some distinction in swimming and in fencing; he studied music; but most of all, he devoted himself to intellectual pursuits. These not only comprised mathematics, chemistry and natural history (what today we should call zoology and botany) but also engineering and political economy. In 1824 he published his only work of significance. This single paper which attracted little attention and was, in fact, virtually forgotten for more than a generation, can now be seen to be a work of genius. It was a small book written in French and entitled 'Reflections on the motive power of heat and on machines fitted to develop that power'. It described in mathematical terms the relation between the amount of heat needed to do a particular amount of work and, contrariwise, the amount of work required to produce the particular amount of heat. This is the so-called 'Carnot cycle' which every engineer today must learn and understand as part of his basic education in thermodynamics. Four years after publishing his book, that is in 1828, Carnot resigned his commission. In 1832, at the age of 36, he contracted cholera and died.

Surely, this simple pathetic narrative contains all that is romantic and poetical about science. Here we have the young man of talent spurred on by his genius, pursuing against discouragement and frustration the noble goal of understanding the mechanism of nature. In doing so he is achieving one of the great aims of the human spirit. Abstract conceptual thought is the one achievement by which man is superior to the beasts. From the beginning of time there have been those who want to know the secrets of the world they live in to find out how birds fly, how they navigate at the due time half-way round the world and back again to the very same nook in the eaves of a barn in the corner of a field from whence they started; why tulips bloom in the spring, dahlias in the summer and Michaelmas daisies in the autumn. But above all this, to relate such abstract conceptions as heat – that mysterious 'fluid' the character of which so bothered the minds of our ancestors – to the work of an engine, of a wheel going round and round and a piston going up and down – to do such was surely the godlike intellect of the human mind at full stretch.

Anyone with a trained mind can recognize the merit of Carnot's achievement as an intellectual feat, but the scientific world was busy with its own affairs and, as is the state of affairs today, made up of people no more perceptive outside their own narrow compass than anybody else; and many years were to pass before Lord Kelvin in 1848, long after poor Carnot was dead, pointed out that Carnot's work besides comprising a fundamental discovery of major theoretical importance also contained the key by which practical treasures of economic value could be unlocked. And Kelvin promptly unlocked one of them – from which it followed that one of the more successful brands of early refrigerators was called the 'Kelvinator'.

There can be no doubts about the greatness of the social changes brought about by the refrigerators, the deep-freeze cabinets, the blast freezers, the tanker-loads of liquid oxygen to fuel the moon rockets and the liquid-nitrogen pizza-freezers that his discoveries made possible. He had, in fact, pointed out that heat – this strange fluence – even though it was a ghost without form, weight or visibility – could be pumped about like water. Each time the little motor turns itself on in the kitchen fridge, it hums an anthem in praise of Nicolas Carnot.

Another example of science triumphant. James Watt was born in Greenock, not so far from Glasgow in Scotland, in the winter of 1736: he represented everything that has, up to now, been recognized as admirable in a scientist. Working as a humble mechanic in Glasgow University, he put to rights an inefficient model of a steam engine, which even when new would never have worked properly. He soon began to *think* about the way water becomes steam, and had the wit to discuss his observations with Dr Black, a first-rate scientist who discovered what is called today the latent heat of steam. Watt recognized its significance and had a flash of intuition, the kind of thing which of all else brings a quiver of pure joy to the true discoverer and entrances us duller souls when we hear of it.

'It was in the Green of Glasgow,' he wrote later. 'I had gone to take a walk on a fine Sabbath afternoon. I had entered the Green by the gate at the foot of Charlotte Street – had passed to the old washing-house. I was thinking upon the engine at the time and had gone so far as the Lord's house when the idea came into my mind that, as steam is an elastic body, it would suck into a vacuum, and if communication were made between the cylinder and the exhausted vessel, it would rush into it, and might there be condensed without cooling the cylinder. I then saw that I must get quit of the condensed steam and injection water . . . Two ways of doing this occurred to me. First, the water might be run off by a descending pipe, if an off-let could be got at a depth of 35 or 36 feet and any air might be extracted by a small pump' – obviously not a particularly practical proposition – 'the second was to make the pump large enough to extract both water and air . . . I had not walked further than the Golf House when the whole thing was arranged in my mind.'

This was but the beginning of the great triumphant march of science, the management of heat, the production of steam, the fabrication of iron and steel to contain them, the control of this potent 'elastic fluid' whereby it could press outwards with the kinetic energy of its pulsating molecules and then, quenched by cold and, like a shorn Samson, deprived of its force by vacuum created in the condenser, caused to pull rather than push – all this was greeted with universal applause. Here was progress unalloyed by doubt. What greater praise could be lavished on triumphant Britain than to call it the workshop of the world! Or if there were doubts, if Wordsworth could point out that 'getting and spending' with the force of the steam engine could lay waste the powers of man to be human, the main current of the times was to praise the achievements of science and the skill of engineers.

But science was perhaps placed on its highest pinnacle as a philosophic principle by which modern man could guide his behaviour by Rudolf Diesel. Born in Paris and educated in Germany where in 1880 he graduated at the head of his class at the Technische Hochschule in Munich, he not only believed that science could be harnessed to do practical things but also that the same process of thinking by which more efficient machines could be developed ought also to be applied to bringing about more efficient social administration.

When he left the Technische Hochschule, he took a job as a salesman in the refrigerator-manufacturing company which his teacher, Carl Linde, with no nonsensical notions about the purity of high scientific thinking and the vulgarity of trade and following the example set by Lord Kelvin had set up. It was while he was employed in this firm (which, in its time, enjoyed the same prestige that manufacturers of lasers and miniature computers enjoy now) that he happened to go to a lecture about the enormous successes achieved by James Watt in improving the performance of steam engines. At this lecture it was proudly announced that because of the innovations and the discoveries of Watt, the efficiency of the steam engine had been increased to the previously unheard-of level of *7 per cent*. Diesel was horrified. Could it be that all the progress for which the great men of the nineteenth century had been applauded had led to the acceptance by men of science of a machine which was 93 per cent *inefficient*? Diesel therefore set himself to design what he called a Rational Engine which would perform with something of the degree of efficiency that Carnot's cycle implied an engine should be capable of.

For the first ten years of his effort he discovered that practice could be more difficult than theory. During this time he worked on an idea that arose from his knowledge of the refrigeration equipment he dealt with in his day-to-day work and tried to construct an engine like a steam engine but one using ammonia instead of steam to make the wheels go round. This is theoretically possible but practically difficult. For example, whereas a slight steam leak does not do much harm, the smallest leak of ammonia brings tears to the eyes of everyone in the workshop. His next idea, though it seemed outrageous at the time he thought it up, turned out to be a brilliant one. It occurred to him that it was a wasteful arrangement to have a fire-box in a steam engine in which coal was burnt to heat the outside of the steel boiler tubes, thence heat the water inside the boiler tubes, turn the water into steam, which then had to be sent through pipes into the cylinders to move the pistons which actually did the work. Diesel's idea was to inject the coal, which would be previously ground up into powder, directly into the cylinders and combust it there where the work was actually to be done, making use of the oxygen in the air which was already present.

He worked out this idea in practice with considerable ingenuity by fixing conical hoppers on top of the cylinders of his engine. These were filled with powdered coal and fitted with rotating plugs designed to screw into the cylinder exactly the right amount of coal dust. But Diesel's most important innovation was to arrange things so that the most efficient conversion of heat into work and work into heat, as calculated by Carnot, would be achieved. Everybody knows that if one pumps up a bicycle tyre, part of the work applied to compress the air and get it into the tyre is converted into heat – and the valve in consequence becomes hot. Diesel's idea was to arrange conditions in his motor so that the pressure of the compressed air should increase to 250 atmospheres, at which point its temperature would have risen to 800°C (that is, 1,470°F). This would be hot enough to set the coal dust on fire and make the engine go without the need for sparking plugs and an ignition system. Diesel calculated that this system would give him an engine with a theoretical thermal efficiency, not of 7 per cent, but of 73 per cent. In actual fact, no engineer living at that time was capable of constructing a machine with cylinders standing up to such a pressure. He had, therefore, to lower his sights. Even so, he was able to design an engine capable of tolerating 90 atmospheres of pressure. This, he calculated, would give an efficiency of 68 per cent. In 1893 he published a book setting out all his ideas under a self-confident title which, when translated from the German, reads something like 'Theory and Construction of a Rational Heat Engine to take the Place of the Steam Engine and of All Presently Known Combustion Engines'.

From then on it was all plain sailing. There was a great deal more work to be done before the efficient Diesel engine we now know emerged. But Rudolf Diesel's position in his own eyes and in those of his contemporaries was secure. His book described the logical intellectual progression by which he had conceived the idea of using Carnot's principles to design a machine far more efficient, in terms of the amount of work produced per unit of fuel consumed. The patent which he was granted in 1892 claimed that the novelty of his machine lay in its being based on Carnot's theory. Surely this was a triumphant victory for knowledge and reason.

In actual fact, although Diesel *had* used his thinking to good effect, his engine was not, by the time it was made to work, an exemplification of Carnot's principles. As so many scientific reports since have shown, a discovery is hardly ever the result of a series of rationally linked intellectual steps. Yet how enjoyable must the times have been when scientists were prepared to believe that they could move so directly from theory to practice. This was the time of Swiss Family Robinson, a book in which, in the periods of activity between prayers and thanksgiving to God, the ridiculous fictional family were building themselves by the pure light of reason rafts which could not possibly have floated; constructing for themselves farm machinery which would never have worked; and domesticating wild animals which could never have been on their island in the first place.

Diesel was, of course, a cut above the Robinsons. His machine did exist and it did work even if its principle was different from what he claimed it to be. And for this we can forgive him for his intellectual self-confidence. Luckily, Rudolf Diesel's last idea – which was just as rational as the thinking which led to his machine – was never accepted by his contemporaries. If, he argued, by taking thought in a rational way a man may make a more efficient machine than a steam engine, surely the same kind of rational mental effort might equally improve the efficiency of a community's government. But let me leave the discussion of this conception to a later chapter.

We look back to the heyday of science and see it applauded as one great success after another for the human intellect. Watt and his tea kettle have an assured place in the hagiography of science side by side with Thomas Edison and his electric lamp bulb – and where would civilization be without that, or for that matter without his gramophone? (However could we decide then who was top of the pops?) And what a backward place the world would be, to be sure, without Mr Biro and his ball-point pen or Mr Clarence Birdseye and his fish-finger.

But even if now we cannot share without question the undoubting praise of their contemporaries, let us not be ungracious to the heroes of those happy times. What fun those early public experiments must have been, with the savant – a lecturer whose every word was attended to with delighted expectancy – describing the chemistry of nitrous oxide, producing some there and then on the lecture-room bench and giving the front row a sniff until he had them rolling about on the floor in stitches: nitrous oxide was not called 'laughing gas' for nothing. Nitrous oxide has since been overtaken by later discoveries; chloroform – once accepted as the safest anaesthetic of all whether for the operating room or as an ingredient for cough mixture – has in our own more timid and doubting age been struck off the list. But the world has indeed benefited from the relief from pain and suffering that the great figures of the last century brought. Our successful modern anaesthetics derived from their work.

Sport, the activity in which prizes are won, possesses certain unquestionable virtues. The people who do it improve in health and strength. Their minds benefit too, freed from the miasma of sloth which inevitably overtakes the sedentary sybarite in slippers sunk deep in his sagging easy chair. There are moral virtues as well. Cricketers do their best to intimidate their opponents, break their ribs and knock out their teeth by bowling a hard leather ball at them as fast as they can, but they do not intend to make away with them. The partisans at a football match do not intend to kill the referee no matter what they shout and, while they may ruin the special trains provided for them, they do not plan to wreck them entirely. Sensible people therefore would support the pursuit of athletic honours in so far as sport makes those who participate healthy when they would otherwise be unhealthy; forbearing when otherwise they would be intolerant; and just, when without commitment to a cause they might well be corrupt.

The great prizes of science are similarly reckoned to be praiseworthy by almost everyone. Life instead of death from infectious diseases; improved strains of plants and animals to bring plenty rather than famine; bridges and roads; the means of communication; textiles and cities – all these are, at first sight, greater and better than the prizes of sport.

Science has won prizes all the way and – who can doubt it? – will go on doing so for ever. At present we are short of energy, but why should we despair. Rather, let us set to work, grit our teeth, keep our eye on the ball, hit it for six, with a straight bat, right across the board, until success comes up trumps and failure is overcome with a left to the jaw. With the triumphs of the steam engine, the Diesel locomotive, the Biro pen and the telephone behind us – not to mention the smelting of aluminium in electric furnaces and the making of steel in Bessemer converters – who can doubt that science has a prize in store for the man or woman who works hard enough to find a substitute for petrol.

The problem is not new nor is it, of course, basically scientific. The scientists beaver away at their atomic piles, solar furnaces, geothermal drills, their windmills and their tidal barrages not because there is any shortage of petrol but merely because the OPEC owners of oil wells have cornered the market and put up the price. There is a well-documented precedent in the winning of a scientific prize under very similar circumstances barely a generation earlier. In the 1920s and 1930s it was not the sheiks of Asia Minor and their friends who had multiplied the price of one of the essential ingredients of the good life (namely petrol) five-fold but the British, whose lively and patriotic young Foreign Secretary, Mr Winston Churchill, made an arrangement with the Dutch to obtain a 'corner' on another equally important commodity (namely rubber) and thus multiplied its world price by an almost exactly equal amount.

For a decade or so anyone who wanted to buy rubber had to pay the price. They did not like it, any more than people in the 1970s liked paying for petrol. They tried to save but without much success. The collection and reprocessing of scrap rubber did not produce any significant quantity nor was a campaign in America to patch and recycle ladies' girdles particularly successful. When, however, in 1942 the Japanese captured the rubber plantations in Malaya and

Indonesia the prize of finding a chemical process for making artificial rubber out of petroleum became worth winning. And by 1944, it had been won. Immense efforts were made, laboratories were established and all the resources of scholarship and research brought to bear. The experience of German chemists faced with the same problem in World War I was used and provided a valuable starting point. The result of this application of chemical science, of engineering and of administrative skills and the wealth to bring them all together was a glittering prize indeed. In 1940, no general-purpose artificial rubber was produced in the whole of the United States; in 1944, the annual production was over 670,000 tons. From then until now, the communities at the forefront of modern civilization can take pride in the knowledge that through their efforts the wheels of every truck and tractor, of each of the myriads of motorcars, of the aeroplanes great and small as they fly up or come down, and of the motorcycles as well – quite apart from American ladies' girdles and hot-water bottles for British beds – have never run short of rubber.

A prize is a prize. In our hearts we would all like to win a £300,000 prize on the football pools even though the more educated among us know that the record of happiness among pools-winners is not encouraging. So let us turn to another prize which scientific knowledge has won for the human race. One of the first comforts a man cast up on a desert island would miss is soap. If he were a well-informed castaway with a smattering of a scientific education, he would soon set to work to make soap. Having taken the advice which the immortal Mrs Beeton in fact never did write in her cookery book – 'first catch your hare' – he would kill an animal, cut away some fat, render it down in a pot and then *saponify* it by stewing it with a watery decoction of wood ashes (this being the aboriginal *potash*).

There are, of course, as the detergent kings had the insight to discover, many things the matter with soap. Its most basic defect stems from the adage about the impossibility of having a cake and eating it. Fat that is used to make soap is thereby unavailable for consumption at the dinner table. In the middle of the 1930s, scientific brainpower produced out of petroleum washing agents with many-fold the cleansing activity of soap. Overnight *alkylbenzene sulphonate* became one of the most treasured of all organic chemicals as the washing-powder giants stood up to do battle for the 'whiter-than-white' market.

This scientific prize has brought about major social changes in the civilized world. Laundries have virtually ceased to exist, nor can widowed washerwomen expect to make a living any more for the modest capital investment in a washtub and a mangle. Busy factories provide well-paid employment for great numbers, who support the economic basis of the community not by taking in each other's washing, but in the manufacture of 'consumer durables', of which domestic washing machines and their companion spindriers are among the most important. These would not exist if science had not first found out how soap works and then shown the way to make detergents – substitutes fifty times better than soap. Nor have the inventors been satisfied merely to discover a basic detergent. One after the other, ways to improve the seminal ABS (as alkylbenzene sulphonate was soon affectionately known) were devised. For more than a decade, of all the bright ideas patented in the United States, most were improved detergents. Some of these possessed built-in ability to cope with hard water, others were designed to deal specifically with one particular type of textile while others yet again were chemically constructed to deal with some specially recalcitrant kind of dirt. Never before had any community been so clean! Detergents designed, not to wash clothes but to cleanse dishes were used to produce mountains of foam in coal mines where they were used to put out fires. Other kinds were actually used to wash coal – a happy term employed by miners to describe the process whereby the coal is freed from scraps of incombustible rock.

All the time, these inventions freed for use as food the 'inedible' fat which before detergents existed was the raw material for soap manufacture. And how valuable these freed supplies of tallow (an inferior grade of suet) turned out to be in this newly sanitized world in which human numbers were growing as never before. Much of the fat was eagerly absorbed by the pet-food manufacturers, whose trade was increasing just as rapidly as that of the washing-powder makers.

Perhaps stranger still in this peculiar race for the perfect detergent, has been the latest development. As the spectrum of detergents for special tasks was unveiled with each modification as intended, washing textiles, steel drillings or coal perfectly, a new challenge faced the scientific competitors. Soap made from fat would after the washing was over as soap-suds go down the drain and into the rivers; there, as food for living organisms, it was consumed to start its perpetual round, as all food does, into the sky and back to the plants to serve as food all over again. But detergents, having started their existence in the deep and rocky entrails of the ground where the oil from which they were made had accumulated for millennia, were only reluctantly consumed and for the most part accumulated as great shabby piles of foam along the river banks. 'Make us a detergent just as good as before,' the chemists were told, 'but one that the miniature zoo of a sewage works can eat. Do this, and there is another prize to be had.'

Success has now been won as it always has been (or nearly always). Instead of starting with petroleum, a hydrocarbon, and making from that a detergent molecule, the winning team started with sugar, a carbohydrate. The world production of sugar is enormous: it is a commodity which, like gold, can be bought and sold, used in manufacture or buried in a bank regardless of how urgently it is needed for filling the cavities in people's teeth. Let not those with a sweet tooth grudge scientists the success they have achieved by the use of sugar in making detergents. Not only are these detergents 'biodegradable' so that they rot, decay and vanish when, like dead leaves, they are swept away. In addition, they overcome what was growing to be one of the significant problems of modern civilization. When motorcars were first designed early in the twentieth century their brasswork and body panels were kept shiny and polished by the loving attention of their owners' servants. As time went by, people had something better to do (we need not at this juncture specify what it was) than polish cars, nor was it necessary that they should: the automatic carwash had been invented. Then it was found that although the detergent mixed with the whirling sprays of water projected onto each car as it passed through the mechanized bath performed its duty admirably on the body-work, the windscreen and the wheels, on the chromium-plated bumpers it fell short of perfection. They were left streaky. The new detergents made from sugar, however, were found to be not only biodegradable but admirably effective in producing bumpers as shiny as the day they were installed. And so another obstacle to the good life was overcome.

There is no need for surprise that in the scale of values the aesthetic quality of chromium-plated bumpers should rank high as a target for intellectual endeavour. Such has always been so. Long ago in the great days of the Roman Empire, knowledge, effort and treasure were expended in the high endeavour, for which no expense was spared, to dye their wool. So costly was the only dye then known by which this could effectively be done that the Emperors alone were able to afford to wear purple – rather as in present times only the Monarch can afford to lock up so much capital in the form of the Crown Jewels in the safe-deposit of the Tower of London. To start with, the dye manufacturers had to know that what they wanted was pus squeezed out of a cyst near the head of one of two kinds of not particularly common shellfish found in the Mediterranean. One of these is now recognized as being *Purpura haemastoma* and the other *Murex brandaris*. According to Pliny, to extract the dye from these creatures one put them in salt and then boiled them for several days. Even then, to get a good colour one had first to coat them with a special liquor extracted from a plant now called *Anchusa tinctoria*.

Today all this trouble is unnecessary. Nor does one need to be a Roman emperor in order to satisfy a taste for purple clothes. Anyone can buy them at Marks and Spencers. The royal Tyrian purple itself, once so laboriously extracted from shellfish, is now identified as 6:6-dibromo-indigotin and even this knowledge is not much of a prize in the ever-victorious forward march of scientific success. It is a sad and curious fact that the purple for which the emperors strove is today merely one in a category of dyestuffs. Yet the discovery of modern dyestuff-chemistry was a major prize for those who won it. In the heady days of scientific enthusiasm of the 1850s, a young student of 18, W. H. Perkin, set by his chemistry teacher to make quinine in the laboratory found that, although he had nothing like quinine in his test tube, what he had got was a very effective purple dye. The whole family then joined in. Young Perkin patented the process he had discovered, his father and his brother put up some money and in no time at all he had not only set up in business as a manufacturer of 'aniline purple' but also opened up a whole new branch of chemistry admirably suited to the genius of both the British and, almost immediately, the Germans. These dyes, designed by the logical processes of chemical experiment, were not derived from the pus of any shellfish, they were made from coal.

But the chemical industry was not merely to acquire great wealth and prosperity from the aesthetic demand for colour – a frivolous quality, it might be argued, of no use or function in the busy hardworking world. Perkin's original (and wholly admirable) attempt to make quinine, led to something more powerful than colour: explosives. The coal-tar chemistry from which the new *diazo* dyes were derived was the very same, slightly modified, from which picric acid and Lyddite were manufactured. After one or two varieties of higher or lower explosive power had been developed came what is today the measure of all explosive worth, the successful basis of every nation's strength, trinitro-toluene, that is, TNT. As we know so well, science does not stand still. When rather later there was urgent need to halt Hitler's Nazi hordes, atomic bombs were developed. It was perhaps bad luck for the Japanese that before the work was done the Nazis were already defeated. But atomic bombs were a major scientific prize nevertheless.

*

Examples of scientific success might continue indefinitely. At every turn people who have wanted to do something have – through science – been able to do it. Successes in medicine come almost at will provided someone is prepared to invest enough effort and money. Poliomyelitis is crippling the children; a remedy is found. Pneumonia comes to kill the old; an antibiotic will stop it. If eight hours is too long for a flight across the Atlantic we will build a plane to do the journey in four hours. Thousands at home want to watch a football match; a television camera allows them to do so – in colour. They would like to see the last goal kicked again – slower; a 'replay' will show them. This is not magic, or self-indulgence, it is science.

Of course, not everything is possible, nor are the public desires entirely clear. No one actually wanted a xerox copying machine or a nylon shirt before these things were invented. But once they were invented, people wanted them. The strange fact is that the products of science by which the twentieth century has come about – electricity, electronics, warmth, light, power, automatic telephones to anywhere on earth, medicine, motorcars, planes, clothes and machines – all the complicated rich and happy today has, in spite of stops and starts, been moving higher and higher up a ladder to ever loftier standards of living (as the quaint phrase has it). And it has all happened without anyone actually planning it. 'Everyone has won,' as the Dodo in *Alice in Wonderland* put it at the end of the Caucus Race, 'and all must have prizes.'

3 This way, Utopia

Everyone knows what an elephant is. At the same time in the scientific age in which we live it becomes more and more difficult to draw up a satisfactory specification acceptable to the Foods and Drugs Administration of the United States, the Food Standards and Labelling Regulations of the Ministry of Food, the British Standards Institute, the Fair Trading Act and the Elephant Breed Society (if there is one) whereby we can be assured that when we are offered an elephant we are really getting one. Inevitably we worry that its quality may not stand up to the appropriate analytical tests, that its health may not be good enough to pass the scrutiny of the Veterinary Authorities of the United Nations or that genetic screening may imply that a *mésalliance* occurred on the borders of Assam when its forebears were engaged in the timber trade. And if the present gives cause for apprehension what are we to think of the future? Should we continue along the road towards consumer protection so impregnable that nobody shall ever be deceived, misled, poisoned or distressed? When that time comes, we may ask ourselves, will anyone ever be able to enjoy an elephant at all?

The technological society has come to be the dominating

culture of our times, towards which every other society, when it calls itself 'developing', hopes to develop; but the members of this society have come to a moment of choice – and the choice is how shall we define an elephant. Either we can strive to define one in more and more refined detail: that it is a creature (a) of the vertebrate variety; (b) big (with a size and weight schedule as laid down); having (c) two floppy ears; (d) two tusks; (e) two eyes; (f) a trunk at the front end; (g) a tail at the back; (h) that it is a mammal; (i) that it has a good memory; (j) a thick skin – and so on with designation of its biochemistry, its blood grouping, its immunology, on and on until no possibility of doubt remains. Or we may choose a quite different approach by saying, as we point at the beast, 'That is an elephant.'

Almost exactly one hundred years ago in 1875, the world was altogether different, a world of horses only just becoming adjusted to railway trains, a world with few telephones and fewer effective drugs, a world of soldiers in red coats and sailors in ships still mainly propelled by sails; and into this world was introduced – a veritable birth-pang of the world to come – the Sale of Food and Drugs Act. The passage of this Act was a sign that the British community had decided to take all due steps, using the powerful engine of science, to define each item of diet, not by saying 'This is a Bath bun and that a sausage,' but by according to each a specification which, as the years have passed by, has grown so intricate and detailed that only with the assistance of a properly qualified analytical chemist can we now know for sure that one is a bun and the other a banger.

To celebrate the passing of a century since the Sale of Food and Drugs Act came into force a meeting was held of all the heads of the official enforcement agencies, the leaders of the scientific community and officials of international and world authorities. So pervasive has science become and so generalized is the philosophy that Man must know what is good, safe, pleasant and nourishing and that no one anywhere shall suffer misfortune, harm, distress or indigestion – that the whole world has come to be concerned in the just composition of the British sausage.

This congregation of expert and responsible persons reviewed the history of the hundred years that has passed. They noted the continuous accretion of legislation and the expanding elaboration of food standards and regulations, one following the other as analytical refinements and the minutiae of nutritional understanding grew and grew. And they then put their minds to forecasting what more could be done in the years ahead to standardize and control the composition of food for the good of society – that great amorphous mass of strangers – so that safety could become safer yet and nourishment more nourishing.

No humane and responsible person would want to live in a community where the food in the shops was dangerous and each man had to employ a taster at each meal to make sure that he was not poisoned. Things were not quite as bad as this in the industrial societies of 1875, but adulteration of food was rampant. Those were the days when, according to the words of a popular song, 'Little drops of water, little grains of sand, make the milkman merry, and the grocer grand.' Something needed to be done to stop the watering of milk, the addition of alum to flour and even so flagrant a malpractice as the use of 'sugar of lead', as lead acetate was then called, to sweeten beer. Progress in analytical chemistry made it possible to police the 1875 Act. It is interesting to find that one of the earliest of the scientific associations, the Society of Public Analysts, was in fact set up in the same year. The regulations were designed to ensure that food on sale should be of 'the nature, substance and quality' demanded by the purchaser. Provided his product came up to this straightforward standard, the merchant was, like any other citizen, a free man innocently selling anything he pleased. Science came into the picture by providing a means of measuring how much butter fat, let us say, a particular jug of milk *did* contain: so that had a purchaser expected to buy milk containing 3½ per cent of fat, and provided a magistrate or a jury thought it reasonable for him to expect this much fat in the milk he bought, it could actually be proved whether the milk purchased had 3½ per cent or only 3¼ per cent of butter fat.

With the passage of time and the advance of science, changes have come between buyers and sellers, and in the expectations – and also, be it said, in the suspicions – of purchasers. The development of food technology in all its modern ramifications and this parallel modification of the philosophical outlook of the citizens of science-based civilizations have led to a radical difference in approach to the assessment of what is understood by good wholesome food. To start with, the analyst's ability to detect smaller and smaller traces of contaminants has enormously increased: figures in terms of parts per billion are commonplace where a few years ago parts per million were all the analyst could expect to be able to measure. And there are certain compounds which a modern chemist, equipped with the great engines of science – the mass spectrometers and machines for determining 'nuclear magnetic resonance' – can detect in a food when only 1 part is present in 1,000,000,000,000 parts of the commodity.

Then again, the much simpler problem of ascertaining whether a medieval taster after having helped himself to the dish prepared for his master had or had not fallen down dead, or even whether a particular group of drinkers had or had not suffered from poisoning by lead acetate immediately after having consumed a few pints of adulterated beer was now replaced by something much more subtle. This was the exceedingly difficult problem of assessing whether, because a proportion of a colony of experimental rats to which the dye, Butter Yellow, had been administered developed cancer of the bladder, people consuming foods coloured with Butter Yellow would within a period of years be likely to suffer also. The difficulty of coming to a sensible decision is rather curiously exemplified by the proviso which has now been quite generally accepted that if people consuming the substance under examination can be shown to have suffered harm, not immediately, but within a year, something should be done. Even if the demonstrable harm only appears after five years, or even ten, the substance under trial could reasonably be condemned. There are instances where blame has been apportioned on evidence only apparent twenty or thirty years after the potentially harmful substance has been on the scene. It is however generally accepted – at least, this is so at the time I write – that if damage can only be expected after seventy years, then the compound – the dye, the 'additive', the slice of bacon, the spoonful of pickles, the pinch of snuff (and all these have been under study) – may be accepted as harmless. This conclusion is based on the proposition that it does not matter being poisoned after one is dead.

The new and fundamental problem arising from the advance of science in the production and manufacture of a food commodity is, however, the virtual impossibility of a normal everyday consumer – whether a professional man or a cheerful housewife, being able to recognize anything as being of the 'nature, quality and substance demanded'. Once upon a time the housewife with her shopping basket and father at the dinner table could recognize a pot of jam or a sausage when they saw either one or the other. Now, in order to appear on a public shelf as 'jam', the comestible claiming to be such must conform to an appropriate chemical composition, must contain a due percentage of designated fruit, be coloured by an approved pigment, be preserved from decay by an acceptable preservative and *not* be preserved by a whole string of usable but not acceptable substances; it may be thickened by this and not thickened by that; the receptacle in which it is presented to the public must be of an appropriate composition closed by a lid or seal acceptable to one or several regulatory authorities, if it is made of metal it must be made of the right metal and not the wrong one, or if it is made of a prohibited metal – for example, steel – it must be covered by a layer of appropriately inert composition. How can an ordinary person know that this is so? I will not enter into the supplementary problem about which, oddly enough, neither the regulatory authorities who protect the welfare of the population nor the scientists and analysts have so far made any worthwhile comment, namely whether the jam is good, nice-tasting jam.

Only if she were accompanied by a retinue made up of a

qualified food analyst, a microbiologist to assess the hygienic virtues of each purchase, a nutritionist to verify the specified content of each of the several vitamins listed on the label and a legal expert to determine whether the legislation laid down by the local authority, the national parliament, the European Economic Community, and the Codex Alimentarius of the Food and Agriculture Organization of the United Nations had been carried out could the conscientious housewife know whether the food she bought at the supermarket really was of the 'nature, substance and quality' demanded.

During the hundred years or so since 1875, the community of Great Britain – as indeed the much wider assemblage of all those who have accepted the potentialities of science to develop technology as one of the main elements in their advancing civilization – comprising the ordinary citizen, the scientist, the diligent guardians of the public good and the food manufacturers themselves have all been carried together along the stream. No longer do any of these believe any more that it is proper – or indeed feasible – for an uninstructed individual by himself, guided only by the light of nature to recognize a loaf of bread or a jar of pickles. Step by step in the pursuit of safety the corporate will of the people, that is to say, the law, has taken upon itself to set up quantitative standards of composition, to lay down detailed specifications for what may appear on the label or in an advertisement. That is why it is suggested that no longer should anyone be permitted to market a long-established comestible at present called a 'digestive biscuit' under that title. It is assumed that since all biscuits marketed by reputable biscuit-makers are capable of being digested, the digestibility of all biscuits may be accepted as part of their inherent properties, so that it would not be fair for one particular kind to claim any special virtue in this area. Only approved colours, preservatives, flavours and other substances, added to improve the texture or retain the moisture – and so properly described as 'additives' – can today be added. And behind all this stands the scientist who alone possesses the skill to determine whether the law has been obeyed and the rules kept and on whose continuous research further refinements in the regulations depend.

Refinements of the current rules are coming as we watch. The artificial colours specially prepared in chemical factories to beautify the food and drink of every advanced technological society are fed first of all to generations of rats which, after living out their lives and raising their young on brightly chromatic diets of many colours are subjected to every kind of post-mortem examination the wit of man can conceive. It has almost come to the stage that if out of a large enough colony of rats, fed from infancy to advanced old age on diets containing more and more of the colour, preservative, humidifier, emulsifier or general improver under test one or two of the animals do *not* develop heart failure, liver disease, kidney breakdown, skin disease or tumours something must be wrong somewhere.

Such artificial colour or chemical additives as survive the modern safety tests – and some do – must be safe indeed. So much so that attention is being turned – as it is only rational (if that is the right word) that it should be – to natural substances which have been, in the innocence of our prescientific hearts, added to various articles of diet. One group of substances which is being subjected to current scrutiny is that used to add flavour. It has long been known that horseradish is a poison and there is a documented case of a pig* which ate almost a pound of horseradish and died. The main reason for not banning horseradish as a relish for roast beef in spite of the comparatively high concentrations of isothiocyanates in it is that it is so excruciatingly painful to poison oneself with it that few people have the fortitude to try. On the other hand, the British Ministry of Agriculture, Fisheries and Food† have thought it worth while to tell the food eaters for whom they are responsible that they consider onions, leeks and garlic; strawberries, lemons and pawpaw; kumquat, shaddock and bergamot; quince, figs and

* Forsyth, A. A., *MAFF Bull. 161* (London), 1954.
† *Rept. on Flavourings in Food*, Fd. Additives and Contaminants Cttee, MAFF (London), 1976.

fennel; litchi, sapodilla, pomegranate and tamarind; bog wortleberries, prunes and parsnips may all be safely used. But they say there are others that introduce an element of risk. 'Do not,' they warn, 'flavour your food with more of the following than the amounts that *you would naturally expect to find*.' This list of flavourings includes cutch, agrimony and silver wattle; grains of paradise, lady's mantle and the rhizomes of great galangal. The lesser galangal can safely be used too, together with hollyhock blossoms, chiretta and cherimoyer. The official list reads like poetry. Turnip, pot marigold, paprika and caraway appear side by side with the exotic sweet vernal grass, alps mugwort and flowers of ylang-ylang (also known as *Cananga odorata*). All of these may be added in moderation to contribute their flavours to the food we safely eat. But what should not be added, as the analytical chemists, the toxicologists and those who conduct the rat autopsies gather their evidence, are roots of the pomegranate, the bulbs of squill, any parts of the lilies of the valley. Nor, they say, should deadly night-shade be used to flavour the food we eat.

The problem is a serious one calling for serious considera-tion, but it is a problem possessing two aspects, only one of which is within the realm of science. This is taken seriously – perhaps too seriously. The scientist can measure the num-ber of rats whose bladders have broken down under a long-continued diet of which a significant fraction has been the sickly-sweet cyclamate. The scientist, however, cannot measure the strength of the desire of the lover of lemonade for sweetness, the legitimate rights of such a lover of lemon-ade to be able to buy his drink cheaply and, having drunk the drink, not put on weight as he – or possibly she – would otherwise do if the sweetness were provided by sugar. Nor can a scientist know more than anyone else whether or not a free-born citizen is prepared to take the risk – surely long odds, at that – of cyclamate poisoning, twenty, thirty, forty years hence or never, or of a reduced lifespan which, with actuarial certainty, is inevitably associated with obesity.

New knowledge should be followed by new precautions if dangers newly identified are worth avoiding. Bacon, a nourishing, traditional food was made as a means for preserving the meat of the pigs slaughtered in the autumn for use throughout the winter. If now, as it appears, the smoked and pickled meat from which the succulent breakfast rasher has for so long been sliced does indeed contain nitrosamines in amounts sufficient to present a risk decades ahead of cancer (unrecognized by our less well-informed and usually less long-lived ancestors) then surely it is sensible to give up our traditional bacon. After all, with our refrigerators at hand, we have other ways of keeping our slaughtered pigs than by hanging up their pickled hams from hooks in the kitchen ceiling. But the balance is a fine one, both from its technical and from its spiritual points of view. The answer to the question asked by Elizabeth Barrett Browning, 'How do I love thee?' is beyond the limits of science whether the interrogation applies to a lovely woman or a slice of bacon. This is the spiritual part of the problem. The technical part is the need to assess the degree of risk arising from consuming the trace concentrations of nitrosamine which, no matter what circumstantial evidence may say, has never been proved to have harmed anybody and compare such circumstantial evidence with the small but certain number of countable cases of botulism which have certainly been found to occur when insufficiently cured bacon is eaten.

Undoubtedly it is sensible to do our best and apply all the scientific knowledge we have to ensure that food is wholesome, uncontaminated by dangerous substances and properly labelled. But there is a question to be asked of those who, looking back after the first hundred years of the Sale of Food and Drugs Act, begin to turn their minds towards what more should be done during the next hundred years that lie ahead, and it is this: have increasing health and happiness *always* followed each successive addition of sophisticated control and regulation? When an American scientist thinks he sees a poison in tuna fish and a Swede smells trouble in imported beef; when a South African researcher finds a toxin in peanuts and all over the place the laboratories cry out against butter, lard, coconut oil, omelettes and Jersey milk; must the whole world take heed if it is to be saved? The unification of the standards of food composition and of the analytical procedures needed to verify their maintenance is already under way – first, throughout the European Economic Community but soon throughout the whole world. Like balloons being blown up to their full size and tension country after country leaves the elementary, undistinguished and varied condition of 'primitive life' and reaches the happy, uniform, money-using status of being fully 'developed'. More than 300 years ago, a Council of wise men was assembled by James I of England to prepare an English version of the Bible, that everyone could read and obey. Today, a body of qualified food scientists drawn from diverse corners of the globe is well advanced in its work of preparing the *Codex Alimentarius*, the world gospel of food composition.

It is right and proper for all these great men, distinguished in science and prominent in government administration – nationally, internationally, and world-wide – to review the progress that has been made; and the end of a century of science applied to the control of food quality, aiming to attain safety so absolute that the innocuousness of eating shall always be assured, is a good time to stop and consider. For example, is the target of perfect quality and absolute safety attainable? And even if it is attainable, is it desirable? Do people fuss and worry about the safety and wholesomeness of their food – as they increasingly do – *in spite of* the devoted efforts of the Food Standards Committee, the Food Additives and Contaminants Committee and the Labelling of Food Regulations, or *because* of them? Will more people be saved by forbidding the sale of bacon and no more men and women of sixty die of cancer forty years or so on from now as a result (a difficult thing to measure, since you cannot both eat your bacon and leave it uneaten); or shall we, by forbidding bacon factories to make bacon, kill some few more people now with botulism?

In the days of Britain's glory as the workshop of the world when Lancashire was alive with steam-driven machinery spinning yarn and weaving cloth for export to every corner of the globe and, particularly, to the multifarious population of India, there was a saying among those whose drive and talent and entrepreneurial skill made an Empire rich: 'Where there's muck there's money.' Within the factories and mills the men and women working there communicated with each other in sign language. They did this because the noise of the looms was so clamorous that they could not hear themselves speak, let alone understand each other. There was noise as well as muck. In the streets of prosperous and – within the sort of limits implicit in Dickens – happy nineteenth-century England there was a din of horses' hooves on cobblestones, the grind and clatter of the iron tyres of carts and carriages, the shouting of hawkers, the cursing of drivers. In the wealthier parts of the town, should someone be lying ill in a front-facing bedroom, it was possible (for a fee) to arrange for straw to be laid down in the street thirty yards or so upstream and downstream of the house to quell the noise.

Today we believe that there is scientific evidence to show that noise can damage the hearing. Or if it does not actually cause any measurable degree of demonstrable harm, it can be argued to cause 'stress' or, at the very least, distress. And this must, in the totally safe world we are aiming at, be harmful. Consequently the criteria within a legislative instrument, the Noise Abatement Act, can be given in scientific units, decibels. Thus, as the level of permitted copper and lead in the safer and safer food diminishes, so does the level of noise in the environment fall. It is not yet obligatory to play Tchaikovsky's 1812 overture *piano*, but the time for this may come. On the other hand, regardless of the increased safety for the musician's hearing, there may come a move to leave loud noise as it is. There is, perhaps, a measure of paradox in the scientific pursuit of noise abatement of factory machinery and the parallel increase in the amplification of the music played, at the special request of the factory operatives, to enhance their enjoyment.

The whole matter is one of balance. But it is a balance best struck by ordinary people on their own, using their own judgement about the kind of life they want to live and the kind of inconveniences they are prepared to put up with.

This means that *not* every judgement can – or should – be left to the scientists. In the predominantly science-oriented atmosphere of the last fifty years or more it has come to be accepted that people do not want to live near their work. It was the people, not the scientists, who originally decided that they would divide their lives into two parts: one part when they were not doing what they called 'work'. The enterprising banker, like the enterprising burglar of W. S. Gilbert's lyric preferred to live well away from his counting-house:

'When the enterprising burglar's not a-burgling,
When the cut-throat isn't occupied in crime,
He loves to hear the little brook a-gurgling,
And listen to the merry village chime.'

This was different from his medieval forebears whose life and work tended to constitute a much more homogeneous whole.

The idea gradually came that 'work' was a fraction of man's life on earth which was a deplorable necessity and, if it could not entirely be avoided, it must at least be carried out discreetly hidden away from where the real living was done. Then the scientists stepped in and, taking up the challenge of making all things absolutely safe, began to set standards of noise exposure which could by objective, measurable criteria be designated as dangerous. Better to travel fifty miles by train to work and fifty miles back again each evening if by doing so one could live out of earshot of potentially hurtful noise.

Subtle indeed is the problem – for noises in the air as much as for putative toxins in food (or tobacco). In a well-run cotton mill the girl operatives, proud of their skill and dexterity, enjoying the company of their fellows, aware of the usefulness of their work, appreciated and properly treated by their managers, could enjoy the purposeful racket of the machinery perhaps even more than their successors enjoy the vacuous clamour of the loudspeakers turned up at their express request – in place of silence. And the sign language the girls themselves invented gave them some of the pure joy that comes from every creative endeavour.

There are two sides, as in cricket or rounders. One side contains serious public servants, members of parliament, local government officers and concerned scientists feeling their way through the impalpable complexities of measurement, trying to determine when the first symptoms of danger appear; from any such danger it is their business (or so they believe) to protect their fellow citizens. The other side is made up of ordinary people. They would never have considered the possibility of danger from noise, any more than from bacon, unless someone had told them about it. And even when they *are* told, if it is their own noise (or their own breakfast) they would still be prepared to take their chances. In the discotheque, on one's own motorbicycle or, if one is older, in one's own motor cruiser or private aeroplane, who cares about noise? If you don't like it, you don't have to go to the dance. Of course, if it is someone else's noise, that is a different thing. The British and the French actually own the Concorde aeroplanes, and individual Frenchmen and Englishmen whose houses are situated along the flight-path may be prepared to claim damages for the harm its noise has done to their property or health. But for an American householder in the United States, the danger he would apprehend and the evidence of harm his scientists would be able to measure would be of a different degree of intensity altogether.

The present Age of Science is a remarkable period of history. In the past, people were well aware of the uncertainty of life. All around them they saw their friends being knocked down by runaway horses, losing their money when their houses burned down, being the victims of arbitrary power or suffering the law's delay. On top of these ills, their children died in infancy, their wives in childbirth and their neighbours from smallpox. Today, none of these occur – or if they do they are infrequent. And so it happens that quite intelligent people – nay more! – particularly earnest people, banded together in the Society for Social Responsibility in Science, are able to set down their aims – aims which they see as attainable – in terms such as these:

The conference is intended to try to establish the outline of the technology necessary if all mankind is to be able to have a good life throughout the twenty-first century in equilibrium ... with the environment. A 'good life' must necessarily include an interesting worth-while job which enables the individual to earn an adequate standard of living for himself or herself and their family while leaving enough leisure, energy and time for self-fulfilment through creative hobbies and sporting, social and cultural activities and access to nature. It must also include freedom from fear, danger, noise, strain and loneliness, (and from) invasion of privacy.*

How strange such a statement will appear to our successors; how bizarre would it have seemed to our pre-scientific ancestors; how odd, indeed, it reads to some of us today. Since the dawn of history every man and woman has sought what seemed to them to be a good life but how few of them have attained it. What can it have been about the science of the otherwise gloomy inhabitants of the 1970s that made a university lecture-theatre-full of scientists believe that it could of itself command a 'good life' for, not just the citizens of comparatively fortunate London, but for *all mankind*. And how odd it reads that the aspirations which science is to bring about, while they include a job (once called a man's life's work) that can be described as 'interesting and worth-while', particularly prescribes a hobby (which, by definition, is not serious work but merely something done to pass the time) which is categorized in much more enthusiastic terms as 'creative' and conducive to 'self-fulfilment'. As one who has always been at pains to avoid the pseudo-pleasures of 'sporting activities', I trust that, unlike the 'compulsory sports' of the traditional English boarding school, it may be permissible for those who so desire to submit a note from home excusing them from this particular item in the programme of the New Jerusalem.

Strangest of all, however, is to find a group of people, even such as believe most fervently in the powers of man as exercised through science, who in an English university of the 1970s could conceive of a life of real matter-of-fact

* Conference at Queen Mary College, University of London, July 1976.

people on the real, down-to-earth terrestrial globe which would be safe from fear, danger, noise, strain and the occasional presence of unwelcome intrusion or the absence of desirable company. For who would enjoy such an unchallenging condition should it ever come about?

The notion that a scientific community could – not by fasting and prayer but by the application of enough of the potent brain-power of science – attain the Holy Grail of safety, purity and 'freedom from fear, danger, noise, strain, loneliness and the invasion of privacy' is, if I am right, the final phase in the period of belief in the omnipotence of man. It is my contention that history is closing the door on that period and that many of the symptoms we see today are but the hinges of history turning.

An alternative aspect of this same strange belief is the assumption that all the disasters that have occurred and all those yet to occur are due to the misguided actions of people who have paid insufficient attention to the word of science. Take the example of carbon dioxide.

Because there are more people on earth than there used to be and all these people are breathing out carbon-dioxide gas, the amount of carbon-dioxide gas in the atmosphere must be increasing. There are, of course, fewer buffaloes in North America and, since a buffalo breathes out a good deal more carbon-dioxide gas than a man, there must have been more of it produced from the two of them in the old days than man alone produces now. This is overshadowed, however, by the carbon dioxide the men produce with their motor-cars, central heating and electricity power stations. Of course, if we go back a long way further than the buffalo days, as long ago as the time when there were more volcanoes than there are now, the amount of carbon-dioxide in the air must have been much more than it is today. People, however, do not commonly bother about this. What they do concern themselves with is the reflection that the increase in human numbers added to the amount of coal and oil burned to produce heat and power will – if the calculations are right – increase by a small but measurable amount the concentration of carbon dioxide in the atmosphere.

Seeking for safety but apprehensive of danger, the scientists started warning the world of unseen dangers ahead. The peculiar form of the disaster they saw and about which they were anxious that busy politicians responsible for the day-to-day running of their community should pass laws and take action was the 'greenhouse effect'. It was argued that a significant increase in the carbon dioxide content of the atmosphere would lead to more of the sun's heat being trapped, and this would make the whole world's climate warmer: ice would melt at the Poles causing a rise in the level of the oceans. For the British community such flooding was estimated that the only safe real estate investment in Edinburgh would be the turret apartment under the flagpole on top of Edinburgh Castle.

But belief in the potency and in the perspicuity of man (and particularly of scientists) is now becoming modified, and the cataclysmic ideas of the immediate past are being modified also. People have begun to notice that some summers are warmer than others regardless of what governments do: it is realized that the greenhouse effect due to carbon dioxide would not be left to operate in isolation and would be influenced by complex turbulences in the atmosphere which, far from being a static covering over our heads, is a complex system of interlocking dynamic phenomena. Furthermore, the carbon dioxide concentration itself is the result of an ever-changing balance made up of the amount that comes into circulation, from whatever source, and the quantity that is removed from the air into the green leaves of all the herbage that grows – also a variable factor.

Those who came to see man the omnipotent controlling at will his estate – the countryside around him, popularly known as the environment – thought he should insist that, not only every step he took by the application of his so-effective scientific understanding be safe, but also the natural situation that exists on earth should be made perfectly safe. In 1975, Professor Richard Scorer, a mathematician from Imperial College, London, stood up as one of the earliest to see, firstly, that the Holy Grail of absolute safety on earth was unattainable; and, secondly, that, powerful though he might be in warming his buildings, curing disease, flying to the moon and inventing hairsprays to keep his ladies' permanently waved hair in order, man was after all a fairly puny creature even in the twentieth-century circumstances he had fashioned for himself. It seemed to Richard Scorer* that it was sensible to take pains to ensure that hairsprays did not damage hair, but it was verging on the ridiculous to believe that the jet from a hairspray in the hand of a modern Cleopatra seated at her dressing-table could not only devastate her Antony, freshly returned from his office desk, but could also destroy millions in far-off lands. Yet a belief had come into currency that the compressed gas, by means of which hairspray droplets were projected from the can, would after its escape eventually rise up and accumulate in the sky: in the upper atmosphere its substance – a compound of chlorine, fluorine and carbon – would be disrupted and the chlorine part would combine with some of the molecules of ozone that exist high up in the firmament.

Ozone has had a varied reputation in the modern history of science. When I was a boy it was reputed to be good for you. The curious smell emerging from the first electric underground railway to be built in London was claimed by the populace to be ozone. Protagonists of underground travel boasted of the therapeutic virtues derived from this ozone, formed, it was believed, by the electric sparks from the travelling shoe that picked up energy from the live rail. Similar virtue was attributed to the breezes blowing in from the sea at various coastal health resorts. 'Ah,' claimed the cognoscenti, sniffing the air, 'it's the ozone that makes a holiday worth while.' It later came to be recognized that the smell they so much savoured was not that of ozone at all but merely decaying seaweed.

Today, the hypothesis most widely accepted by the science-oriented, firm in their pursuit of safety and their belief in the efficacy of science to attain it, is that ozone in the upper atmosphere forms an absorbent screen: that otherwise so strong would be the sun's ultra-violet energy striking people's skin that it would induce cancer. This

* *New Scientist*, 66, 702, 1975.

hypothesis is a complicated one. Too much ultra-violet radiation can undoubtedly be harmful to pink-skinned peoples like ourselves – who have with the passing of the millennia evolved to a present state which enables us with luck to live to about seventy in the chillier parts of the earth. If there were less ozone we should become more sunburnt than we are – and this might be bad for us. On the other hand, the evolutionary inheritance of brown and black-skinned people has fitted them to live nearer the equator where there is more ultra-violet irradiation. They might legitimately claim that to have to live in Chicago, Detroit and Wolverhampton is bad for them because there is *not enough* ultra-violet in the sunshine of those regions. So far they have not pressed the government or the United Nations to take active steps to *reduce* the ozone layer in those places, but they might.

Having reached the improbable conclusion that hairspray and pressurized cans of liquid starch and paint affect the composition of the sky high up over the whole globe of the earth – as was the conclusion of intellectual worriers in the early 1970s – it was hard to swallow the new information that came to hand in the late 1970s. In 1976, Alan Eggleton, Tony Cox and Dick Derwent of the Medical Sciences Division of the UK Atomic Energy Research Establishment reported the results of detailed studies of what really does take place in the stratosphere.* As the winds swirl around in the upper atmosphere, as the sun shines and the earth spins the things that happen are quite complicated. The tiny injection of the propellant gas from the hairsprays may in fact cause several effects, but it was not entirely clear from the new measurements whether any of them are significant at all. The refined calculations of Drs Eggleton, Cox and Derwent suggested that the total tonnage of propellant was minuscule. American researchers had found that the natural concentration in the upper atmosphere of hydrochloric acid – which was thought to be the dangerous product of the ozone reactions with aerosols – was twice as high during one season

* *New Scientist*, 70, 402, 1976.

of the year as in the other. No divergence as big as this had ever been attributed to aerosols.

Scientific measurements need to be checked and checked again and fitted within the context to which they apply. Down among the parts per hundred million and the parts per billion (as Americans designate a thousand million) the results may wobble about somewhat. There are means of calculating the 'statistical significance' of any particular measurement, but a prudent scientist will constantly reassess the conditions under which his measurements have been made. One can easily calculate that the results of a survey indicate that 3.76 per cent more voters express a preference for left-wing rather than right-wing candidates for political office, or that the blood pressure of a particular group of people is 6.37 per cent higher than the norm – and that both these figures are statistically significant. The new wisdom – the heresy gradually gaining strength in the general ambience of scientific orthodoxy – is that both these figures (in spite of their accurately calculated statistical significance) can be wrong.

For example, if the poll of people's voting intention was taken when opinion was changing rapidly, or immediately prior to the disclosure of some monstrous scandal about the principal minister, even the most diligent scientist might have failed to notice it; whereas the shrewd, non-scientific man of affairs might pick up a sniff or a hint of impending trouble from the gossip of the pub, the stock exchange or the queue at the supermarket check-out. In his laboratory, the experienced researcher also has a nose for true values. He can spot contaminants in his reagents, a bad connection in a recording galvanometer, or infection among experimental animals, all of which may render his measurements worthless. It is just because he *is* so incompetent in the affairs of real life that his proposals about hairspray and cyclamates are often so naive.

In the old days, before the belief in the absolute power of *Homo scientificus*, there were prudent people who set up machinery to ensure that factories were run as sensibly as could be arranged. In pragmatic Britain they had the Alkali

Inspectorate, the Noise Abatement Act, the Clean Air Act and corresponding regulations; some of the mess caused by the ebullient days of the Industrial Revolution of the eighteenth and nineteenth centuries was put right without bringing the business of the community to a standstill. The new hazards following utilization of nuclear fusion, aeronautics, rocketry and the higher biochemistry as applied to therapeutics and entomology needed new measures. The new ethos of the American Environmental Protection Agency and the world view of the United Nations Environment Programme (UNEP) implies a belief in Absolute Safety on earth through Science and Absolute Democracy – the apotheosis of man – and represents something quite different.

Yet even though the science of geology demands the removal of Naples to a safe distance from Mount Vesuvius and of San Francisco from the San Andreas Fault, the citizens of both are content to leave their fates in the hands of Providence and stay where they are. To lean against the post office in Aberdeen is to expose oneself to a level of radio-activity from the native granite greater than that permitted by United Nations' safety standards – yet Aberdonians stubbornly persist in staying in Aberdeen. Some people even continue to smoke, just as in the days of that most altruistic and scientifically unassailable public-health measure, the Volstead Act, some American citizens continued to drink.

Perhaps after all people would rather sniff dangerous air, eat potentially toxic foods – potatoes with their poisonous solanine, cabbage with goitrogens, plums containing prussic acid, rhubarb full of oxalates and cassava high in cyanogens. They go on riding hazardous motorcycles, playing rugby football and risking the germs inescapable in kissing. Science would protect us from all these; democracy, the concentrated essence of the wisdom of man, would undoubtedly vote for the pursuit of a life of total innocuousness, just as Americans voted to stop their own drinking. Yet other targets for life on earth retain considerable resiliency, even in our most scientific age. We go our own ways. After all, the old-fashioned Kingdom of Heaven was not a democracy.

4 The stammerer

One of the most interesting, and perhaps important, questions that people ponder in modern technological societies is whether practising scientists have a special moral responsibility for what they are doing and for the uses that are made of the new natural knowledge which their researches throw up. There are two schools of thought. One subscribes to the view that a scientist, like a cutler, *is* responsible for what people do with what he provides: if he supplies the knowledge to enable his employer to produce sharp, rustless knives and those who buy the knives use them to stab their wives, the cutler must take the blame – even though the knife he made might just as well have been employed to chop up the carrots for the Irish stew.

The other school of thought is that the discovery of the internal structure of the nucleus of the uranium atom, for example, enriches mankind's store of knowledge: it opens the possibility of dating archaeological specimens, of watching the mechanism of the cells of the living body, and of creating available power to supply heat and light. The power may be used to run industries by which goods to provide a fuller life can be manufactured – or to make terrible weapons by which the enemies of civilization can be deterred.

But whichever school of thought is right, both hold to one principle: in view of the diversity of people who become scientists, each with individual aptitudes and weaknesses, scientists owe as their first moral duty to the community of which they are fallible members the duty to publish their findings. They must tell their fellow citizens what they have discovered and, as far as they themselves can see, what the implications of this new knowledge may be. It is then for the community as a whole, through whatever political mechanism it may select for its own organization, to decide what to do with the knowledge, or indeed whether to do anything with it at all.

Moses, like so many great men – and like all too many scientists – was a poor speaker. 'O my Lord,' he is quoted as saying (Exodus 4: 10), 'I am not eloquent ... but I am slow of speech and of tongue.' Anyone who has been to a meeting of scientists will be all too well aware that the worst defect they tend to suffer from is not only being inaudible and difficult to understand but having a pernicious tendency to go on much too long. The solution hit upon all those years ago in biblical times is still valid today. It was to invent 'the media'. Even in those days, however, there was tension: the anger of the Lord was, it will be remembered, kindled against Moses for not making a better effort to do his own public speaking and for making it necessary to put up Aaron the Levite. But the arrangement was made, and as the Lord said, 'I know that *he* can speak well ... and he shall be thy spokesman unto the people.' And so the system was established.

Ordinary citizens sometimes behave as if the marvels of science – the conveniences and the horrors alike – came into being by magic. They possess no clear idea of how science achieves its results, nor how it comes upon the knowledge by which practical things are achieved. Least of all are they clear about the factors upon which the issues they debate so emotionally are based. It is for these reasons that I, for one, believe that the *highest* ethical duty – I had almost written the *only* one – laid upon the scientist as such (he shares, of course, the other diverse duties of a good man and a good citizen with all his neighbours) is to make clear what he is up to so that all the rest can understand. But does the scientist carry out his special moral duty of speaking out? And, if he does, do his fellow citizens listen?

Civilized man, with all his cleverness, refined and stimulated as it is by an expensive education, is deaf. A supersonic aircraft flies by and a great and strong wind breaks the windows. And after the wind, an earthquake produced by an underground nuclear test. And after the earthquake a fire, as the chemical works at Flixborough blow up. And after the fire the small voice of a scientist murmurs in stilted jargon an account of his researches shortly to be published in the scientific 'literature'. Even if he were not deaf, modern man would understand scarcely a word of it.

It is paradoxical to look back to the time when science – still called natural philosophy – was not involved in the practical business of making goods, running the trains or relaying the spoken word and the animated face to the four corners of the earth. Then there were no trains to run and broadcasting had not been invented. In those days scientists wrote in plain speech like other educated men. Darwin's *Origin of Species*, presenting evidence upon which a new conception of man's beginnings was closely argued, was a best seller from the moment of its publication. Today, a scientist who is a specialist in biology is almost incomprehensible to, say, a chemist or a geologist (not to mention to an educated man or woman who happens not to be a scientist at all). And the more devotedly science is studied, the more numerous the divisions within one discipline become; paint chemists and dye chemists; men who study the chemistry of the aroma of port wine; those who investigate the chemical structure of plastic sheets or lubricating oil; less and less easily can they understand each other, even though they are all chemists. Not only do the different kinds of scientist fail to grasp the meaning of the multitudinous scraps of science with which their colleagues are concerned, but they and their fellow citizens who live outside the magic circle of the cult find it still more difficult to notice the ground swell of the advancing tide of scientific knowledge.

There are some men of science who try to tell what they see ahead. Among them are those who announce a 'breakthrough'. Sadly, this is often little more than the break of a prisoner who burrows out of his cell only to find that his tunnel has come up not in the freedom of the outdoors – but in the next cell. Hard on the heels of these misguided prophets come the journalists – glib Aarons to a man – interpreters whose business it is to search for the truth in the confused turmoil of daily life. Having found it, they make it interesting to those for whom it is written, so that they shall at least read it. Is it fair to arraign the journalists? Their livelihood depends on engaging the interest of people who for their own pleasure and edification buy newspapers or watch the box. The first step must be to persuade them not to put down their papers or switch off. Do journalists really set out to titillate their audience by plucking a single sensational fragment out of some new scientific discovery which they themselves only half understand? Or is it that if they went on and on talking about the identification of a homologous series of fluorobenzyloximines out of phosphatic jurassic soil specimens, they would lose the attention of their audience?

But if the voice of science fails sometimes to get across because the professional communicators need to be lively and not too long-winded, are the scientists themselves blameless? To a scientist, 'the literature' has come to mean, not great thoughts stretching out in search of truth well-written in well-chosen words, but a tortured text, laboriously phrased to pass the scrutiny of dried-up referees selected by the editors of scientific and technical journals.

The voice of science – the stammering utterances of the new Moses who, carrying his potent rod by which the Nile can veritably be turned to blood and plagues of locusts encouraged or destroyed – is not only muted within the intricacies of its jargon but it is often solemn as well. But where, it may be asked, are the Bernard Levins of science?

They are all too few. The most penetrating modern instrument for reaching men's minds is television, but it took some time for producers to learn how to present the facts, and the implications of the facts in an entertaining and engaging way. Today broadcasters, who once upon a time were respectful to science, are following the precedent set by journalists: first of all worming out of scientists what they are trying to say and, secondly, establishing what the scientific facts really are. They are no longer prepared to believe that what scientists say must be taken as gospel; they investigate them as ruthlessly when they talk about the supposed doings in their laboratories as they would were they reporting Sir Jasper's describing what happened in the library on the night of the murder.

And it is every bit as important for the responsible citizen to know what goes on in the laboratory and what it portends. The 'official statement' appearing in the scholastic jargon of the scientific literature is no longer good enough.

There are signs that scientists themselves, awakened to discover that talented writers and broadcasters can interest them (and entertain them meanwhile) in topics about which they are as ignorant as their neighbours, are beginning to call for a louder and more entertaining voice even in the solemn and obscure pages of the scientific journals.

There are occasions when the scientists, even when they do try to speak out in their thin small voices, find that their fellow citizens do not want to hear, even the journalists who are supposed to be trained diggers for the buried truffles of truth, or the responsible people with political power and social influence. It was unpalatable to the British ear, particularly if the ear was Scottish, to accept that Herr Rudolf Diesel had made a better job of converting the heat of combustion of fuel into work than had Mr James Watt. Perhaps the ordinary man might argue that there is no need to dwell on a disagreeable scientific truth such as this now that the implications of the truth have been accepted and the steam locomotives have all been converted into diesels. (There is a moral in the fact that it is no longer necessary to use the capital letter of Herr Rudolf's surname.) But this may not be so. Current affairs may indeed once more be demonstrating how important it is for the scientist to be able to speak and the ordinary citizen to listen.

Those who first developed nuclear power stations have a right to be proud of their scientific knowledge and technical skill. They and the laity among whom they live are prepared to accept uranium-fuelled power stations. Even if they do not exactly like having them, people are reasonably content to enjoy the benefits they bring in houses heated, goods manufactured and numbers employed. Yet when the scientists and engineers whisper that the efficiency with which the fuel – after its laborious discovery, extraction and purification – is converted into heat is only perhaps 9 per cent, nobody wants to listen; least of all do they want to know that in a 'breeder reactor', the plutonium fuel is converted into heat with about ten times greater efficiency.

Conscience always seems to be such a bore when it whispers to our better judgement that the course of action we ought to take is one that involves trouble or even hardship. This is one of the factors that makes democracy such a disagreeable as well as unreasonable system for operating our affairs. It is irrational because it depends on the mass of the citizens – the lynching gang, the rampaging football crowd, King Mob (as it was once known) – to arrive at a sensible solution to whatever public problem may be involved. And it is disagreeable: regardless of the fact that we all like to find someone else to blame – the government, the multinationals, or even 'science' – for any mistake of ill-judgement, we citizens of democratic communities know that we only have ourselves to blame.

To bring a man of the 1870s into the 1970s would be to demonstrate the greatness of the power and influence of science on the lives of ordinary people. A hundred years is short enough a period in human history. Two hundred years is little longer, yet the citizens of the United States in 1976 dramatically drew attention to the changes that had taken place in material conditions during the short life of that republic. There is good reason to shout from the housetops about what science has done: the comforts it has brought;

the command over events; the understanding of life, death and the nature of the universe; to tell everyone what this facet of the genius of man has accomplished. Yet somehow, when the scientists do shout, what they have to say all too often sounds empty or even downright silly.

There are two main reasons for this. First, the important innovations that bring about substantial changes in the social scene are usually built not on one discovery but on a combination of discoveries, some big, some small, which only exercise a significant influence when they are all put together. The history of science is littered with discoveries made, as the saying goes, 'before their time'; that is to say, before the other supporting bits of knowledge had come to hand by which the potential of the first could be actualized. But the second reason why the voice of science is so often muffled, unhelpful or even downright misleading is because science, though it can be argued to be something special in the way of thinking, is only part of the general flow of scholarship. Scientific knowledge and discovery do not stand on their own but bring their influence to bear only in combination with knowledge and thinking which are not scientific at all. A scientist, who, outside his science, is ill-informed or ignorant, can be extraordinarily ineffective when he chooses to raise his voice. His wild wood notes may easily be out of tune with the rest of the chorus.

In rich America, as well as in other lands where the better-class members of society are not only rich but rather shame-faced about it, the voice of science can while being right in detail be wrong in every other way. 'White bread is poison,' the cry goes up; 'science says so.' 'Cornflakes and Crackly-Crunchies are "empty calories". Down with them!'

What science says – or to put the matter another way what the facts are – concerns vitamins of the B-group, namely, thiamine, riboflavin, and niacin. While white bread contains all three of these compounds, brown bread contains slightly more. This circumstance, though true, is only of practical interest to anyone other than scientists if people eating white bread are actually going to suffer by consuming less of these three ingredients. Until science began to talk about them, nobody else had heard of thiamine, riboflavin or niacin. But it needs matter-of-fact Aaron, a man of the world, to construe such scientific statements into commonsense terms. It is to the public misfortune that, perhaps because of the dazzling glitter of science, he does not always succeed in doing so. To take another example for a moment, science shows that flannel shirts made out of wool are more effective insulators than similar shirts made out of nylon, but the wool people and the nylon manufacturers do not bother to publicize the insulation effectivenes of their respective textiles in terms of kilojoules per square centimetre or other incomprehensible terms; ordinary people, while well aware that there must be an optimum requirement for warmth, as there is for food, can judge for themselves whether one shirt is warmer than another. So, since no one in Great Britain and precious few of the people in America who fuss about white and brown bread have ever seen anyone suffering from beri-beri, ariboflavinosis or pellagra due to eating too little of the B-vitamins, far less too much white bread, it is little wonder that they have no means of judging whether or not the authors of the so-called 'scientific' books on nutrition and the writers of articles in the glossy magazines are talking sense or not.

For years, 'science' said – or was quoted as saying and scientists made no denial – that fish and chips, at that time the food of the poor, was unwholesome and nutritionally deplorable. Though this is now known to be nonsense (and, fish having become expensive, fish and chips have almost attained the dignity of being recognized as an exotic delicacy), the voice of science has never been heard to recant.

'Science says' that as people get older, their judgement of distance deteriorates and their reactions become slower. 'Let us,' conclude those who listen to such sayings, 'keep such older people off the roads lest as they drive their cars, they run into and kill each other, occupy expensive accommodation among the scientific instruments of our hospitals and damage, or at least choke up, our expensive and scientifically engineered motorways.' Science is right to say that the reactions of old men and women are slower than those of the young, for this is true. Yet the interpretation is misguided; it is the young men and women with quick eyes and instant reflexes who have the accidents, because regardless of their superior equipment they drive more recklessly.

Who, then, is to blame when the impression gains ground that it is the reedy voice of science saying that white bread is bad for you and that young people are safe drivers? Is it the fault of the ordinary people for not listening more carefully to what the scientists say? Or is it the fault of the scientists for, at best, not speaking clearly or, at worst, for not telling the truth? Or – dare it be hinted – for not understanding the implications of the measurements they make and the phenomena they use.

The story of vitamin B (as it was once called – thiamine as it is now) is a wonderful one. I have told it before but I hope I will be forgiven for telling it again in this context. Eijkman, a doctor in what was once (how long ago it all seems now!) the Dutch East Indies was trying to find out the cause of beri-beri, then a crippling disease of poor labourers in that part of the world. What could it be that caused them to lose the use of their limbs and waste away or, on the other hand, develop dropsy? In those days so many diseases had been shown to be due to the attack of some potent malignant micro-organism. Yet none could be found to be the cause of this. In the middle of his serious medical studies – so the story is told – it was found by one of those boringly conscientious administrators one finds everywhere that Dr Eijkman's servant had been obtaining rice from the military stores to feed his chickens. Yet – or so the administrator ruled – Eijkman was a civilian. It was, therefore, quite wrong that his civilian chickens should be fed on military rice. The supply was consequently cut off and Eijkman's servant was compelled to buy rice from the local shop. Not long afterwards the chickens became ill and began to tumble about as they lost control of their legs. Furthermore, they developed a twitch and some of them twisted their necks almost permanently the wrong way round.

Eijkman's brilliant deduction was that, firstly, the symptoms were in many ways similar to those of beri-beri, secondly, that the bought rice was the high-quality polished rice much like the rice we make rice pudding out of. The military rice was rough and brownish. Eijkman concluded that there must therefore be something of *positive* virtue in the rough brownish material (the polishings) which were polished off the brown rice in converting it into white rice. It was lack of this beneficial substance in white rice polishings that caused chickens to develop their beri-beri-like disease and *not* the presence of some harmful malevolent bacteria.

Before long all this was proved. Scientific researchers all over the place extracted rice polishings, and by one chemical treatment after another succeeded in isolating the pure substance, the vitamin *thiamine*, from it. In due course its molecular structure was identified so that it could be manufactured (as it is today) by the ton.

Knowledge of the primary cause of beri-beri and the means for its prevention and cure was a major prize of understanding. But perhaps of even greater delight was the success that soon followed, based on work carried out at Oxford by R. A. Peters using pigeons. This work elucidated exactly what function thiamine plays in the biochemical combustion by which life is kept going. Naturally, once all this was known it was not long before estimates were drawn up of how much thiamine people of different sizes, running their bodily engines slowly or fast, needed to keep the system going sweetly. Soon, however, the eyes of the scientists turned away to other things, to problems yet unravelled. 'Yes, yes,' they snapped, 'a 70-kilo man does need his thiamine. Of course, of course!' If they had been pressed, they might also have answered the other question, 'If the body of a man or a woman *is* running sweetly, does it not mean that he or she is getting all the thiamine necessary?' with a yes. But they seldom *were* asked this question, and the good scientists who had themselves made the discoveries were asked even less frequently to consider where the thiamine of rich men, poor men, beggarmen and thieves came from. Nor did their business bring them in contact with the poor. Yet poverty restricts the variety of foods that anyone can afford to buy and, at the ultimate, the amount of what they can afford within the conflicting demands of clothes and shoes, cigarettes and beer, rent and light, television sets and fares; all of which make their demands on the resources of the poor as insistently as on those of the rich.

So, to sum up, when the voice of science is called to pronounce on the nutritional welfare of the poor, which is a factor in their health – in itself is part of their happiness, or lack of it – that confident voice of science may bear false witness. 'Science says that what is needed is more thiamine.' 'Science says that a slice of brown bread contains more thiamine than a slice of white bread.' So be it, these statements may be true. But while they may be true, they may also be unhelpful. The poor may indeed need more thiamine (particularly if they live in Egypt or the Philippines) but they may also need more urgently many other things. And if they live in the kinds of countries where the voice of science is most persistent and attacks on white bread – or sugar, or fish and chips, or additives, food colours, and sweetened breakfast foods – most frequent, the advice which 'science' is alleged to give may be bad advice for two reasons. Firstly, the poor who are so altruistically recommended to change their diet may not like the new regimen they are strongly advised to eat; secondly, if the 'improved' diet is more expensive their well-being may be diminished rather than increased.

Is it the duty of a scientist, besides telling of his discoveries in as clear and explicit terms as he can, also to be entertaining? The answer is surely yes. A good historian is entertaining; Keynes and Galbraith entertain as well as instruct when they deploy their new ideas about economics. Why then should a scientist, almost alone among scholars, be at best incomprehensible and at worst tedious? The beautiful atom, spinning through space with its electrons flying round it like a ballet skirt, is entertaining. Sir Hermann Bondi, so well does he speak, succeeded in making the derivation of the seminal equation, $E=Mc^2$, by which the energy residing in the core of the matter is related to the mass of matter within which it exists, entertaining even though the intellectual concepts upon which its truth is based are complex. The story of Galileo, laboriously clambering to the top of the Leaning Tower of Pisa with two balls, one heavier than another, is bizarre to the verge of farce. That all bodies, large or small, massy or minute, fall when they are dropped at the same rate *provided there is nothing to hamper their doing so* is one of the fundamental attributes of earthly existence. Curious, unexpected, comfortable in its logical inevitability, it cannot fail to entertain the educated mind. The anecdote about Galileo is entertaining in a different way. If it ever took place, how disappointed Galileo must have been when his two balls did *not* reach the ground together. This was not because the elegant entertaining hypothesis was wrong but because when things are dropped from the Leaning Tower of Pisa, or from anything else in the open air, for that matter, there always *is* something to impede their fall, namely the air. But the story is full of romance (Galileo and the lovely tower so many years ago) and of humour (all those misguided intellectuals who have pompously asserted that what did not happen did). It almost warrants the attention of a Jules Verne to invent a hero who would surround the Tower by a circular wall, apply a lid hermetically sealed at the top and pump out the air so as to allow experimenters wearing breathing apparatus to go inside and drop the two balls off the upper gallery to fall to the ground simultaneously in a vacuum.

To describe the voice of science when speaking at its best as being of its very nature entertaining does not imply that science is a joke. On the contrary, what it has to say is of great importance and seriousness.

Professor Alfven, a Swedish cosmologist, has drawn together some of the links by which the voice of science is tied to all those other voices of different kinds of people perceiving the universe in which they live. In the beginning, said Alfven, people gained their understanding of the universe which lay beyond the horizon by interrogating travellers who had been there; and they found out what the world was like before they were born by asking their grandparents what things were like when *they* were young. Before long, ideas about distant lands which lay beyond the experience of travellers and distant times past beyond the memories of the oldest inhabitants were reinforced by myths and fables about the origin of the earth and the gods that once lived – such gods bearing many peculiar ungodlike attributes remarkably like those of the people who invented them. It was then only a short step to postulate that these gods, turning up in a manner not easy to explain in an eternal but chaotic universe, *created* the world and its immediate environment at a time varying in antiquity according to whether the particular gods had themselves been invented in the Middle East or India. But wherever the ideas came from, there was common agreement that since the universe *had* been created by gods there must be a sublime order in its basic structure.

Pythagoras not only invented a system of mathematics but also a system of religion. According to the Pythagoreans the most 'perfect' of all geometrical figures is the circle and the most 'perfect' of all solid bodies is the sphere. It therefore seemed to them axiomatic that when the gods created the system we live in they created the earth a perfect sphere around which revolved a number of inviolable crystal spheres on which the stars were stuck so that they too could revolve, amid peals and symphonies of sweet music, in perfect circles. And because divine forces can only make perfect systems, all these moving things rotated at a steady uniform speed. Observations seemed in the main to confirm these ideas so that they could be claimed just as well for the science of their age as current beliefs are claimed for ours. The most distant stars – which were stuck on to the sphere which was nearest to God, moved round the sky with almost perfect regularity. Nearer objects, such as planets, were obviously contaminated by disturbing influences which explained the way they wandered about the sky and sometimes actually went backwards. And in medieval times St Thomas Aquinas, like an Einstein of his day, modernized the whole idea by introducing the notion that the start of the universe had been an act of creation in which the entire thing had been created out of nothing.

Although it was beautiful and elegant to be able to describe a universe in terms of perfect circles and uniform motion, so remarkably related to the mathematics of music, there was a difference in the speech of science based on the unchangeable myths of theory compared with the scientific voice of the seventeenth century spoken through the mouths of Newton, Kepler, Galileo and Brahe which says that celestial mechanics does not in fact obey the theoretical rules of divine inspiration. And for 300 years this has been the way that the voice of science has spoken. It has tried to be inspired commonsense, and if some of the topics that science has talked about have been closed to ordinary people because they have not been able to see the bacteria or the stars with their unaided eyes this was a small matter. There have always been microscopes and telescopes of increasing power and sophistication to look down for those who wanted to see for themselves.

The new oddity about the voice of science in our own times came, according to Professor Alfven, with Einstein. As a new Moses, the mathematics in which his general theory of relativity was couched was obscure, often demanding four-dimensional space (which while perfectly possible in mathematics is impossible in terms of ordinary reality). Consequently the Aarons who wrote the popular books read by ordinary people and ordinary astronomers, no longer explained what the universe was really like but merely dished out Tablets of Stone which read: Believe this! Most modern citizens, whether they have been trained as physicists and mathematicians or not, admire Einstein not because he made clear to them the mechanism of the universe but because he relieved them of the need even to try to understand what kind of a universe it is they live in.

The sheer joy of indulging in the intellectual gymnastics of the higher mathematics and the intellectual concept of four-dimensional space began so completely to intoxicate those few gymnasts who were capable of taking part in the performance that they began to forget that, elegant and mutually coherent as the exercises were, they were *theories*; they were not based on any more observation of what was happening in the real world outside than what had supported the earlier Ptolemaic universe where all the orbits were perfect circles, the heavenly bodies perfect spheres, and the speeding stars unvaryingly uniform. Once before, ordinary people – or at least, ordinary scientists who found themselves somehow or other outside the boundaries of accepted belief – had found themselves compelled to contradict what the voice of science believed the universe to be like. The heavenly bodies do not describe perfect circles, nor are they themselves as round as ball bearings in spite of the mathematical logic that insisted that they were. Today, the mathematical logic of Einstein's equations allows them to be solved in a way which would conform to the universe being in a state of expansion.

This is a splendid belief for modern man. And if we cannot understand the rarefied voices of the Einsteins, Gamovs and Hubbles on the mountain-tops of theoretical physics, the words of Aaron – of Abbé Lemaître, and of a multitude of popularizers – were clear indeed. In the beginning (or at least, within a moment of two of the beginning) 'all mass in the universe' – the mountains and seas, the great plains and Arctic wastes, the sun, moon and all the stars, every one of them spread throughout the vast distances of the galaxies – all this mass was concentrated into 'a very small sphere' – a pin's head, a speck of dust. This is the myth of modern man, the fairy tale of the generation that begot the supermarket, the Kojak culture of screaming tyres and hijacked Boeing 707s, the travelling hotel lobbies through which grimly smiling hostesses struggle with trolleys of prefabricated food.

This extraordinary myth of the whole universe created out of nothing as a single grain of primordial sand – what Abbé Lemaître called *L'Atome Primitive*, inconceivably massy and unbelievably hot – was not a conception of some romantic story teller, a Grimm or a Hans Andersen, it did not arise from the primordial fables of ancient times. Not at all. It purported to be the voice of True Science. Yet it demanded a god to have created this concentrated speck of dust so that from it the universe we now know could expand, proliferate, blossom and be fruitful. After all, there has to be a Chinaman to make the little folded paper pellet from which the flowers, leaves, stems and branches spread when it is dropped into a glass of water.

Strangest of all is the growing evidence which suggests that, just as was the Ptolemaic universe befôre, the universe described by the high priests of science in the mysterious language of higher mathematics only intelligible to the selected initiates, may be no more firmly based on knowledge of what the real universe is like. The modern conception may be mythical too. Professor Alfven has itemized a number of the beliefs of the day which, ingenious and theoretically satisfying as they may be, do not describe reality. For example, looking back through time with their radio and x-ray telescopes, theory demanded that an early 'black-body' temperature of 50°K should be observed – but it is not. Nor do the theoretical calculators account for the proportions of all the diverse elemental stuffs that actually exist in the solid substance of the stars. Another discordant fact which grates on the clear voice of scientific theory is the unevenness of the real universe: had everything spread out from a single exploding point, this would not have been expected.

The theoretical calculations require that the universe should be of a certain density but observation can only find about one hundredth of what should be there: 99 per cent is missing. Even observations which are said to show that the whole universe, exploded from its original primordial atom, is still expanding are not uniformly convincing. The commonsense idea that the earth is flat, based, no doubt, on the observation that once you smooth out the bumps, a billiard ball does not roll off it, had to be abandoned when more perceptive observations led to the inescapable conclusion that it was not flat at all but spherical. It is paradoxical to find that nowadays, because the serene voice of science tells us that the universe is an expanding sphere, a burst of outward-rushing shell splinters, the rival observation that shows it to be a flat disc of flat galaxies does not receive much mention.

Every subject of inquiry and learning possesses its own intricacies. To write the narrative of history may require a knowledge of languages, diligence and persistence in discovering the records and documents of times past, judgement and skill in assessing evidence of events and places and the veracity of individuals long dead. In their libraries and in the specialist conferences where they exchange their findings and beliefs, historians can be as incomprehensible as any scientists. Yet an educated layman can expect to understand what happened in the past, nor do historical scholars feel that they demean themselves in writing and speaking clearly and entertainingly to their non-professional brethren. Gibbon's *Decline and Fall of the Roman Empire* is ideal reading for a non-specialist castaway shipwrecked on a desert island, and few people of any sensitivity can fail to fall under the delightful spell of Pepys, whose world has been so richly illuminated in recent years by the historical and literary scholarship of his editors. Because a scientist uses special tools, an electron microscope, let us say, the language of chemical symbols and mathematics of which the ordinary citizen is ignorant, it is more difficult for him to speak to his neighbours. More difficult but not less important. Yet all is not well with the voice of science. It can be blamed for timidity, secretive reticence and downright obscurity; for only partially informing (if indeed it can be said to inform at all) the public who install the automatic telephones, the computer and the xerox copying machines; it can equally be blamed for not shouting about the things that science *cannot* do.

There is an old chestnut about a wife, leaning out of the car window to plead with a policeman to excuse her husband who has run into the back of the vehicle in front, knocked a man off his bicycle, driven through a red light and landed up with his front wheels on the pavement, on the grounds that he was drunk. Today, there are the wives of science, devout in their faith in the absolute veracity of their calling, who believe, with the true devotion of disciples, that intemperance is a state of biochemical maladjustment and that to be drunk is not a vice, a sin or an example of known fallibility for which a man can be held responsible. Instead it is a symptom of disease for which a scientific remedy can be sought. Wickedness, the cool, self-assured voice of science has been known to say (speaking in the dialect of psychology, anthropology, or biochemistry), no longer exists. It is merely a product of the environmental pressures. The wicked man, as he once used to be called, can therefore no longer be held responsible for his actions. Do we not have a Minister of the Environment whose Department can put things right? And if deviant activities (as it is only polite nowadays to call what once used to be known as crimes) are not solely due to environment, they stem from failings in breeding. But since science has unravelled the chemical molecule of heredity, DNA, the so-called 'double helix', so that bright young students can write down its formula, and 'genetic engineering' is all the rage, it will not be long before we have a Ministry of Breeding to back up the Ministry of the Environment. The government can then take steps to ensure that, just as the Ministry of Agriculture draws up regulations to ensure that farm livestock shall be of sufficiently good breeding to produce Grade A milk guaranteed to do us good, the Ministry of Breeding (I forbear to give it an alternative designation) will take responsibility for ensuring that the State shall only produce citizens with the appropriate genetic make-up to guarantee good behaviour.

Faced with a financial expert, a political expert or an expert on horse racing, an educated citizen knows enough to exercise his own judgement. Only thus can a reasonable society exist. It is even more important for the same ordinary educated man to possess the rudimentary knowledge and judgement to be able to deal with the scientific expert on the same basis. The voice of the high priest is all very well, but sensible conversation between sensible people is better, whether they are scientists or butchers, bakers or trades-union shop stewards. We ought to talk over the statements of science with some care. They are rather important and it would be a shame if they were wrong and we never suspected it.

5 A dinner of herbs

Chemistry, the science of the composition of matter, can obviously from its very definition, be taken to comprise every material stuff on earth – and, for that matter, in the heavens as well. Although this gives the chemist an almost limitless remit and allows him legitimately to view virtually everything he sees, whether it is animal, vegetable or mineral, as a combination of mixed chemicals, there are in fact limits beyond which the rule of chemistry no longer applies. Food, its composition, what we think of it, what function it performs in the body, and to what extent it does people good – all these things can be seen to lie along a peculiarly ill-devised frontier; here the realm of chemistry, which, of course, is one of the founding States of the Commonwealth of Science, abuts on an alien kingdom.

During the last century, the chemical approach to life and living, to the composition of the special fuels that keep the human engine going, to the substance of the bricks and mortar, the flesh, blood, bones and sinews of the corporal frame – such studies have proved to be useful to the understanding of the intricate way the animal organism (within which I am including the human organism) operates. And by knowing what goes on, it has proved possible to improve the operation of man and his animals, to prevent things going wrong and to remedy the situation if, in spite of everything, they did go wrong. Antoine Lavoisier was a clever young Frenchman whose father unfortunately bought him a minor title of nobility: unfortunate because when the People's Republic took over in the 1790s they executed him – I wonder what will happen to holders of the OBE when our revolution really gets going. Anyway, Antoine Lavoisier made one of the great discoveries of science when, after reviewing the results of his experiments he was forced to the conclusion '*La vie est une fonction chimique*' (life is a chemical process).

So it is, indeed. A chemist can follow a piece of dry toast from the moment it becomes crunched between the molars. The molecules of starch, mingled with crispy dextrins (provided the toast has not been burnt to black charcoal in the making) suffer their first chemical degradation from the ptyalin of the saliva (spit, to the general public). Down among the more powerful enzymes, what remains of the starch does not stand a chance. The chain-mail mesh of glucose units is untangled and broken apart into little bits which soak into the circulating currents of blood. Chemistry, chemistry. The atoms of carbon and the atoms of hydrogen peel off and reacting with bits of oxygen, some already there, some breathed in from the air, the slow burning which is the process of life fizzles gently, continuously, controllably. Chemical power makes a muscle twitch, a quick smile, a thought, a memory. Or, the fire-box opens up, the chemistry roars, and the great thigh muscles contract as the sixteen-stone man rushes along the street to catch his bus. From toast to glucose, from glucose, linked to phosphorus and unlinked again, to pyruvic acid, succinic acid, lactic acid and a dozen more – the chemistry of life's slow fire proceeds.

And then there is the protein that everyone talks about, the stuff leather, hair, skin, silk, feathers and roast beef are mainly made of, as well as the stiffening in the bread that was turned into toast. 'Body-building', they call it, and so it is, if you happen to be a child that needs to grow. But father does not use much of his interior's chemical ability to build body out of his slice of toast. It's energy to catch that bus he's after. Into the 'fire-box' goes the protein, to be chemically combusted into the exhaust gases he breathes out. The subtle chemistry rejects as unusable the nitrogen part of the protein, harmlessly, ingeniously, if inelegantly, as the compound urea in urine. Bit by bit, the chemical story has been elucidated. The day's activities have been measured in terms of units of energy.

An engineer can measure the energy used up in a machine to turn the wheels and work the pistons. It makes no difference what the machine is doing. It may be working the propellers of a ship or turning the wheels of a motorcar. To the engineer the problem is one of assessing the chemical energy in the fuel and then taking measurements of those parts of the total dissipated and lost as heat and expended in overcoming friction, and comparing them with what remains available to do the useful work of turning the wheels or thrusting the ship forward. His researches may lead to the discovery of an antiknock agent to direct more of the fuel's potential power to making the machinery go. He can properly delight in a modified design for the steam turbine that moves the ship's propellers round. All this is his business and his delight. Alec Issigonis puts the engine of the mini-car sideways on; James Watt provides a separate condenser to improve the efficiency of the steam locomotive.

It is equally possible to study, and to achieve useful results by doing so, the chemistry of life without necessarily giving thought to the purpose of the life that is to be lived or the destination of the creature that lives it. Of course, a farmer feeding a pig knows exactly what he is doing and what the purpose of the pig's life is. From two identical animals he can attain two different results. The pig's food, like that of a child on its way to being an adult, needs to contain the chemical compounds to supply energy, protein, a dozen vitamins and as many mineral ingredients. Yet by the way he regulates the quantities he provides when the pig is young and during the different stages of its short life, the skilled farmer can so regulate its growth that he can either produce a bacon pig, long and flat, that when put through the slicer gives a plentiful supply of rashers; or if he is aiming at sausages, he can feed it differently so that it grows up rotund and spherical. Scientific knowledge, much of it provided by Professor John Hammond, FRS, of Cambridge University, tells him that the legs of a young pig, regardless of the way the animal is fed, have a strong inborn tendency to grow. If in its early weeks of life, he provides only a low level of nutrition, a spare, thin, leggy animal will result. If he feeds it, as mothers feed their young when anxious to produce winners at the village baby show, at a high level of nutrition he will find himself with a fat little pig. But nutritional science enables the pig feeder to exercise a more subtle control over the conformation of the animals under his care. A low level of nutrition in the early weeks of its career followed by a high level later on – more calories, more protein, more vitamins of various sorts, and more calcium, iron and other mineral elements – will produce a pig with a different shape than one fed first at a high level and later on at a low level. As a little pig grows bigger, two waves of growth start to flow, one going upwards from its ankles and the other backwards from its waist. With judicious management these two waves can be made to coalesce at the loin and thus contribute to the prosperity of the farmer anxious to market prime loins of pork. A beast started at a low level of feeding followed, let us say, by an equally low level during the middle period of its life (we could call this a *low-low* pig), has a shape that is different from that of a *high-high* pig; so the farmer has a considerable measure of control over what eventually walks out of the sty on its way to meet its destiny.

For a pig, the destiny – after all this exercise of nutritional science – is comparatively limited. There is bacon, pork, sausages and, if fate is kind, a happy, if eventually somewhat abruptly terminated, family life. The destiny of man is wider. But it is not of man's destiny that the food scientist thinks; or, if he does, he thinks only of that destiny in terms of physical efficiency. It is generally assumed that the final objective of science is to produce from each foetus, the largest, strongest, most muscular, bouncy human being possible. The target, the perfect man, the ideal of scientific research is taken to be the female Olympic tennis champion or the male weight-lifter flexing his muscles on the cover of the health-and-strength magazines. It is as if the farmers all decided that the perfect pig was the rotund provider of the maximum number of pork sausages. In human terms, the sophisticated, long-sided bacon pig, its tissues elegantly streaked with a fine tracery of delicate fat does not exist. The whole edifice of nutritional science is built on chemistry. Roast beef and boiled carrots do not exist to a scientist as roast beef and boiled carrots. Mostly they are water, both the carrots and the beef – simple H_2O. Then there is so much fat and so much protein of a special amino-acid content, a little sugar, some starch and a string of substances in between, the hemi-celluloses and the glyco-proteins. There are a few milligrams of iron and rather more of calcium, some magnesium and phosphorus. It is no accident that phosphorus is an ingredient of match-heads and serves to make them flare up when struck. It does much the same in the slow combustion of life when the oxygen from breath strikes sparks (in a cool, continuous way) with the tinder of glucose circulating round the body as fuel to make it go. Vitamins of all sorts, each with its own known chemistry, join in the processes of biochemistry.

The food chemist, looking at roast beef and Yorkshire pudding or Dover sole with his steely eye can see the ingredients in them that make the body machines of the people who eat them operate – the calories, the calcium; the fat and the folic acid. Folic acid is a minor vitamin that keeps blood the way blood ought to be; without it the blood may run down into a state of anaemia. The keen-eyed chemist sees it because it is almost always there, which is why very few even of the people who *do* have anaemia have it because they don't eat enough folic acid. But even after he has interpreted a tasty meal into terms of protein and potassium, B-vitamins and carbon and all the rest, the food scientist identifies another rag-bag of chemical odds and ends. These are the compounds that contribute aroma and colour. If the scientist's particular interest is health and nutrition he pays little attention to these. He will perhaps like to know what this chemical configuration is, to make sure that they are not likely to do the eaters who eat them any harm. For instance, as I mentioned earlier, recent research implies that one or two of the compounds that are produced by the chemical reactions that take place when one fries a rasher of bacon might, if consumed in sufficient amount and for sufficiently long, cause cancer. Regardless of such happiness as may come from the lovely smell of frying bacon penetrating under the bedroom door and, more persuasively than the hoarse clarion of the alarm clock, driving the sluggard from his bed, the scientific message is to forbid the citizen to eat such bacon. Mostly, however, to the food chemist, the compounds that make food smell, taste and shine in colours of the rainbow from the plate, are of perfunctory interest.

The machinery of life goes not a whit more smoothly whether food is red or green. And as for taste, so minute are the amounts of the compounds that contribute the delicate flavour of prime beef (compared with mince) and caviar (compared with any old roe), that they affect the workings of the tissues not at all. Indeed, the nuance of toast and the delicacy of a fresh fried potato chip owe much of their attraction to palate-stimulating compounds produced when amino-acids combine with sugars. The amount of such combination is minute in terms of ounces and grams but, such as it is, it subtracts from the total of protein and starch. If the amount was worth bothering about, the hard-nosed nutritional scientist would be against it, if he thought about the matter at all. Yet people who are seriously concerned with the food business are well aware that the colour of food, its aroma and flavour are important. Just as in the car business the colour of the upholstery, the shininess of the paintwork, the convenient position of the glove compartment, even the lovely smell of leather and the hushed reverberation like that in a cathedral of the slam of the car door – these are the real criteria which decide whether people consider cars of superior quality and worth paying really high prices for.

The knowledge of the chemistry of the body's needs and of the components of food is valuable for the attainment of the good life. Indeed, it may be essential. Children cannot be saved from the crippling deformities of rickets until the knowledge of vitamin D has been elucidated and the community has so arranged its affairs as to ensure a supply of vitamin D-activity to the children who need it. A healthy and effective population is better assured in wartime by a nation possessing the scientific understanding to know what to put into Woolton Pie and how much Snoek or Iceland salted dried cod to import.* But mention of these delicacies inevitably raises doubts in the minds of those old enough to remember what they tasted like as to whether the chemical science of nutrition, complete and remarkable though it may be, is the whole truth when it comes to good eating and the good health that goes with it. Carry an unconscious casualty in from the street and keep him going for six months on liquid nourishment containing every necessary vitamin from A to B_{12} injected through a tube and the nutritional scientist will rightly earn the everlasting gratitude of the patient's family. But as soon as the experts' efforts are rewarded with success, the injured man restored to consciousness and the tubes removed from his system, he will start complaining about the hospital food.

This brings us to the 'Marie-Lloyd syndrome' of which the motto is 'A little of what you fancy does you good' or, to put the same principle in classical terms, *de gustibus non est disputandum*. In spite of the non-scientific origins of both these aphorisms, scientists as they advance into their new enlightenment are beginning to have more and more to say about them. In this part of the world we do not fancy dog-flesh or horsemeat and even in my own lifetime the taste for bloaters – strongly salted pickled herring – and pease-

* Woolton Pie was a dubiously attractive mixture available to the British in World War II; it was named after the then Minister of Food, Lord Woolton, who made provision for the supply of its ingredients. Snoek was a fish of ambiguous culinary virtues of which there was apparently an oversupply during the 1940s. When shortages were eased it never appeared again on the British market.

pudding, popular when I was at school, have dwindled. These, it might be said, are cultural changes which fall rather within the less exact science (if it is a science) of anthropology rather than within the more precise sciences of nutrition or food chemistry. On the other hand, the diverse dietary intakes of Jack and Mrs Sprat, whose contrasting attitudes to fat will certainly have been reflected in their calorie intakes and will almost certainly have affected their susceptibility to heart disease, probably derived from physiological variability.

There are thus at least three sets of factors which affect our state of nutrition. The first is chemical: how many micrograms of vitamin B_{12} do we obtain from our homogenized beefburgers and how many grams of polyunsaturated fatty acids should there be in the special soft margarine we obtain from the healthfood shop? The second is cultural: would we really rather die than eat the domestic cats and dogs we so affectionately stroke and pat? The manufacture of canned, vitaminized, semi-moist pet food is today a major industry using large amounts of nourishing food which we ourselves could perfectly well eat if we chose to do so. Indeed, reverting to my own youth, I still look back with nostalgic pleasure at the baked sheep's heart which we boys used to consider particularly tasty in our school dinners as the first Wednesday of each month came round. Now, heart is never seen as an item in a human menu but only occurs as part of dogs' dinners.

The third set of factors is more complex and subtle than either of the other two, because it comes from the subtle borderland that lies between science and non-science. On the one hand is man, the combination of biochemical mechanisms kept going, not by rich banquets or frugal repasts shared with trusty comrades under the stars, but by the metabolism of carbohydrates of diverse sorts, proteins in their almost infinite variety and fats, saturated, unsaturated in modest measure or polyunsaturated; the entire mechanism kept going by the necessary intake of sodium, potassium, phosphorus, iron, copper, vanadium, selenium – the list stretches out like poetry. But on the other hand is this same man who, while continuing to be this same mechanism of biochemical reactors, is at the same time a living, thinking, feeling person influenced by joy and fear, enthusiasm and depression, and, most potently of all, by love.

When scientists really did think of themselves as natural philosophers, serious experiments were done to see whether love could be measured in the laboratory. This was in the early days of electricity and there seemed to be no good reason to doubt that the delightful shiver a young man feels merely by approaching his hand to touch the hand of the lady he loves could be measured just as an electric current can – by an appropriate instrument of the same kind. This was, I repeat, in those pioneering times of science before the boundaries of its estate had been marked out. Since then the white lines have been drawn and, though science has been extended to comprise much, love has been excluded from the territory covered by its title deeds. This may not, however, be so as we come to adjust our ideas.

It is, of course, untrue to assert that one can live on love. Just the same, it can be demonstrated even by the strict unemotional methods of science that love can in fact make a contribution to the nutrition that seems so certain in the nutritional textbooks. 'A moderately active man,' the textbook may say, 'expends 2,750 calories of energy in each 24 hours of his life here on earth.' Surely that is precise enough for anyone? If one turns over the page, or looks lower down the tables of figures, it will not be long before the next item of information appears. 'Should the moderately active man change his habits and become an active man,' the text would run, 'he will then expend in each 24 hours of sleeping, eating, dressing, undressing, going to work, coming home again, stroking the cat, kissing his wife and doing his active work 3,025 calories of energy.' A textbook, however, is different from a novel, or an autobiography, a diary or a record of actual events. The moderately active man is not you, neither is the active man me. Both of them are abstractions, they are notional or ideal creatures without any personal existence. The figures in the tables of 'Recommended nutrient intake' are average figures, not intended to refer to any particular person. If one searches back behind the textbook, behind the standard tables, one comes to the reports of real scientists actually measuring the breath of real individuals; weighing them, measuring them, trying to persuade them to submit to the experiments on hand and come to the laboratory before breakfast so that their metabolism can be measured. There are to be found, if one digs deep enough into the tattered notebooks in which the actual figures were written down (quite a different kind of account from what one reads when the results are written up), there may be found the names of the people involved, the trouble they gave, the doubts as to whether they did or did not skip out for a sly bottle of beer. Sometimes there is even the expression of annoyance at one of them (or even, bother it, two) dropping out of the research because they wanted to marry each other. The average may have shown a consumption of 2,750 calories, or 3,025, but among the individuals performing approximately uniform tasks many may have been getting along perfectly well on 1,850 calories and others eating their way through 3,650 calories-worth of meals.

There is nothing surprising about this to the working scientist. He knows that animals living on 'free range' as people do, and of mixed hybrid stock, as the mixed-up populations of even the most noisily patriotic communities tend to be, do vary one from another. After all, that is what sex is all about. It allows the new-born individual to take after his mother *and* his father. And not only to inherit abilities – and weaknesses – from these two. The combination of genes from as we say, the two *sides* of the family are exactly parallel to the combinations and permutations involved in winning the pools. Occasionally, 'eight draws' come up and the 'winner' turns out to be a Shakespeare, that is someone very different indeed from the average, what the nutritional scientist so charmingly calls 'the mean 70-kilo reference man'.

Good scientists have always known that individuals vary. Professor Tremolières, working in Paris, was continually pointing out that, useful though average estimates may be in planning the diets of nations, the best way to find out whether one particular man (or woman) was obtaining his dietary requirements was to ask him 'How do you feel?' Dr Widdowson in Cambridge drew attention too to the variability between people, whether full grown or only babies. Professor Jean Mayer of Harvard studied the energy expended (in the precise terms of calories, of course) of different girls playing tennis: one girl, hopping and skipping, bent double, keyed up, on watch, tense, alert – not actually doing anything but waiting to receive service; her opponent, standing still, waiting patiently for the ball to arrive and hitting it back calmly when it came. If such are the real happenings on a school tennis court, what is the textbook to do to enlighten the earnest reader seeking guidance about the 'energy demands' of tennis? Obviously, the best answer to this question is to collect together as many experimental results as come to hand, study the statistical variability which the physiological differences between one individual and another render inevitable and, using the average values in the light of such understanding, attempt to elucidate factors which do in truth affect what happens.

There is a widely held unscientific belief embraced by many people from periods of history long antecedent to the age of science in which we now live that love possesses an ability to sustain the human frame. While studying the food needs of rats in the traditional statistical way, Dr T. Levine came across evidence which seemed to provide dry, exact – or, at least, reasonably exact – scientific evidence in support of this old belief. His experiments involved the study of two groups of rats, all cousins. Both groups were composed of young rats of exactly comparable ages, evenly distributed between the two sexes. The two groups were housed under identical conditions; each animal was in what, had they been university students, would have been called a study-bedroom – but since they were rats these were more prosaically described as cages. The rooms in which the cages were kept were air-conditioned. Daylight was kept out to avoid its falling unequally on the members of either group. Instead, lights were switched on at appropriate times to represent day and switched off at a predetermined time to provide a measured similitude of night. The diet, designed to provide all the nutrients the metabolism of rat could require, was also the same for the two groups.

One thing only made the lives of the rats in group A different from those of the rats in group B: for group B all the necessary housekeeping, the removal of droppings fallen below the wire bottom of the cages, the provision of fresh water and the supply of each day's provision of the chemically perfect diet was done in silence with deft scientific efficiency; for group A, the girls looking after the animals were told that while maintaining their high scientific standards of precision they should unbend and show their love and affection for the rodent patients under their care. While doing the necessary chores, they were told, they should talk to each rat as they cleaned out its cage, should stroke it, fondle it and call it pet names. At the end of the trial, these rats – upon whom baby-talk, stroking and affection had been lavished – were found to have grown bigger than the others. So nutrition, it would seem, is not a simple biochemical process. The chemical composition of the food the rats ate had something to do with their growth and health; indeed, it had a lot to do with their wellbeing. The results of two generations of brilliant, scientific study are not to be ignored: exclude vitamin A from the mixture and the rats cannot thrive; cut back on the vitamin B1 and they develop polyneuritis; fail to add vitamin D in their youth and they will get rickets. But in spite of the general validity of Lavoisier's dictum that 'life is a chemical process', the dictum is not the whole truth: love them, play with them a little and show them that they are not forgotten and they will thrive better than they do when they are ignored.

Stockmen have known about this for years, but scientists are only now beginning to learn the same lesson. In the experimental farms near Ottawa in Canada were to be seen a year or so ago two groups of young cattle, both housed in similar covered yards. Every animal in each group was provided with a perfectly balanced ration in exact proportion to its weight, in its own named feeding bowl. The beasts of group B were tied up, each one opposite its own feeding place; those in group A were allowed to wander loose. The animals in A group were also provided with individual mangers containing their measured rations of feed, and each manger was covered by a hinged metal lid. No other beast than the one for which it was intended could get at the food, but when the owner came to eat, it bent its head down; a magnetic key, hung round its neck on a leather strap, operated the mechanism, the lid moved aside and the animal ate.

By using this device, called a Callan-Broadbent gate, it was possible to regulate the amount and composition of each animal's ration to provide its exact needs according to its weight, its state of growth, and activity, without denying the beast its liberty. Liberty and social intercourse are as important in their way to cattle as they are to people. Cattle have their friends; they also recognize their superiors and take appropriate steps to keep those they see as their inferiors in their proper place – just like people. If the results of the feeding trials can be taken as evidence, it seems that they are happy to live in an organized, stable, social system. And because they are happy in such a social environment, they do well. In short, the animals in the Canadian feeding trial that were allowed the freedom to set up a proper social organization, grew and throve better than the animals which, while receiving a diet every bit as ample and well chosen as that given to the others, were prevented from working out their social priorities. Here again, it would seem, is a demonstration that the chemical composition of a diet by which the relative amounts of all the different nutrients in it can be assayed is not the sole determining factor in the benefit the creatures draw from it.

Perhaps the most precise measurement of the dietetic influence of love was that which Dr Elsie Widdowson stumbled upon during a study of the nutrition of children – orphans in Wuppertal in Germany immediately after the last war. The experiment actually turned out to be back to front because it showed the effect not of love but of its opposite, unfeeling administrative efficiency applied to a community 'for its own good'. How many generations of children must down the ages have exerted all their emotional powers to counterbalance the biochemical and nutritional virtues of the lovely rice pudding, cold mutton fat, sour stewed rhubarb (and perhaps even workhouse Christmas pudding) administered to them for purely altruistic reasons – but without love?

The Wuppertal experiment was carried out like this. In 1945, immediately after World War II food was comparatively scarce in Germany. Dr Widdowson and her colleagues had under their care a number of children who were orphans of war. They were divided into groups and housed in separate pavilions. The diet they were given was composed of such supplies as were available combined together to provide as satisfactory a regimen as possible. That a tolerably adequate supply of protein and fat, vitamins A, the B-group, C, D and all the rest and of calcium, iron, phosphorus and iodine together with other necessary minerals was provided was demonstrated by the fact that over the first six months of their surveillance, as they were weighed, measured and examined week by week, their health remained good and their growth and development, all things considered, were satisfactory. It is true that the rate of growth of certain of the groups of children was faster than that of others even when every allowance was made to cancel out such possible differences as those of age.

At the end of twenty-six weeks of the study, the food supplies in the district having become more ample, it was decided to see what happened when a measured food supplement was provided for certain selected groups of the children. In other words, some of them were given more to eat. This was done and the trial continued for another six months.

When Dr Widdowson first had an opportunity to consider the results of the trial as a whole she was puzzled. For example, there was one group of children who had grown at a steady rate for the first six months but who when given the extra food – which, it can be said, they ate up as children should – continued growing at the same rate but no faster. On the other hand, a second group whose rate of growth had been steady enough, although it had not attracted any particular attention as being faster or slower than might have been expected, showed an abrupt improvement as soon as the second six months of the trial began. What was peculiar about this was that they had not been given any of the extra food and their rations were, therefore, exactly the same for the second twenty-six weeks of the study as they had been for the first. Only the behaviour of a third group of the children could be thought of as normal. They thrived for the first part of the study and when they were given the extra rations they grew faster still.

Studying carefully these results and – most important of all – thinking over the details of the work, Dr Widdowson hit on the answer. At the six-month break point when the extra food supplies were made available to some of the groups of orphans but not to others, a number of purely administrative changes were made. Some of the people in charge of the children were moved to other work. Among these was a particularly devoted and efficient matron. This lady was conscientious to a fault. Among her peccadilloes – exercised, I may say, solely in the interest of efficiency, discipline and accuracy – was a habit of giving instructions to the group in her charge in great detail, with force and decision, and at mealtimes! In short, she tended to scold the children when she had them all together round the table. They did not like it. Furthermore, they did not like her. So she had succeeded in surrounding her charges in a layer – not of love – but of hate. We can now see what happened. With the first group, during the first part of the trial when there was perhaps only marginally enough food, they were at least able to eat in harmony and peace; when extra food was provided its positive nutritional effect was cancelled out by the negative nutritional influence of the un-love – or even hate – caused by the scolding matron who had now been put in charge. The second group – that showed a spurt in growth rate even though their diet remained the same – had during the first six months been under the yoke of the pernickety matron; now they were relieved of this incubus during the second half of the trial. Only for the third group of children could it be said that their rate of growth both when fed the earlier rations and when, later on, they were supplied more amply was explained by the body-is-a-steam-engine metaphor. That is, the extra input gave extra performance.

Dr Widdowson wrote a long, detailed, technical report to the Medical Research Council describing what had happened to the orphans; what the vitamins they had eaten in their cabbage did for their health and the calcium in their milk ration did for their bones; how many calories they acquired before their rations were supplemented and how many they obtained afterwards, and so on. Having written all this down as a good scientist should and, of course, having set down in detail everything that had been discovered to flow from the phenomenon of the scolding matron, Dr Widdowson could not refrain even in the austere pages of an official report from quoting, to illustrate a proper appreciation of nutritional science, from the Book of Proverbs, chapter 15, verse 17: 'Better is a dinner of herbs where love is, than a fatted ox and hatred therewith.'

The concentrated biochemical knowledge upon which the wisdom of the accumulated textbooks of nutrition is based is epitomized in tables of recommended nutritional allowances put forward by the prophets and seers, the experts and pundits of the Food and Agriculture Organization of the United Nations, the US National Research Council, the UK Department of Health and Social Security and many more from other places all over the world. When all is right and the boy or girl, man or woman actually eating the food of the chemical composition so delineated is an 'average' boy or girl, man or woman – a truly democratic creature falling statistically in the middle of bigger boys or girls, livelier or more torpid ones – the accumulated wisdom of science epitomized in the figures given in the tables will produce a perfect specimen. Or, as we must now agree, will do so if love is added to the mixture as well.

There are, however, different standards of perfection. Insurance companies, for example, have compiled immense quantities of statistics relating height and weight and medical conditions – blood pressure, heart function, sugar content of the urine, muscle tone and many other measurements made by their medical examiners – to longevity. The biggest and best collections of the sizes of children at various ages have been collected by life insurance agents so that the companies shall know who is most likely to die and who to live and for how long. Only thus can they know for how long they can expect to receive their premiums year by year before they must anticipate, with the mathematical certainty of the Recording Angel, that the Great Reaper must appear and demand of them the sum insured. They know, these grave actuaries of death, that when a cohort of men are *x* pounds overweight, their life will end the sooner. Equally are they aware that the perfect life on earth, the life statistically certain to stave off death the longest and ensure the receipt by the company of every premium due, even to the last, is that of the perfect man, tall, well-grown yet abstemious, never to be found in a sports car, climbing a mountain or in a bull ring, a non-smoker, not too fat, of phlegmatic temperament, preferably in Holy Orders, but not a doctor or a missionary. This is the ideal man, every vitamin in place, living a life without risk or stress. Insurance companies take an interest in the physiologically perfect man and woman, the product of the ideal diet providing optimum nutrition, for understandable business reasons.

It is a striking phenomenon of the modern technological age in which improvements, modifications and changes are expected each year in the products of manufacturing technology that large numbers of people take as much interest in their own nutritional status. It is almost as if they expect science to be able to turn them too into an improved new-year's model of a man or woman. To this end they undertake to lose weight – slimming is what the stout middle-aged women and the chunky girls call it – or, sometimes, to gain it. Such people spend a fortune on vitamin pills without any thought to whether their plethoric diet is deficient in one vitamin or another and certainly without any clear idea of their showing real symptoms of vitamin deficiency, of the scurvy which afflicted the sailors living on salt pork as they rounded the Horn, or pellagra from which dispossessed Spanish peasants of a century ago took to their beds, blotch-marked by the sun, and died. The perfect man or woman, those whose attention in the affluent age of the technological society is primarily directed inwards towards themselves, is the possessor of specially designed shoes, fitted with arch supports, and of aerated under-vests; they consume health-salts by night and pure lemon juice in the morning; their weight is exactly that specified in the health-and-beauty manuals; their skin is anointed with skin food; their head is free from dandruff – that dire disease, so widespread in its incidence, so limited in its mortality; and they spend one weekend in four at a health farm directed by a registered physiotherapist.

The third widely accepted image of the perfect human being, enjoying the perfect health which the nutritionally exact diet is designed to ensure, is that set by the organizers of the contests to select Miss World or Mr Universe. All those curves and bulges, those quivering thighs and bulging muscles, those vacuous smiles or – for the men – self-satisfied glares, do these indicate the perfect specimens which we all aspire to be?

Science can say much, but it is incapable of defining the target at which all scientific knowledge is aimed. 'A good diet,' the physiologist might say, 'is one capable of permitting those who eat it to achieve their optimum physiological potential.' But if the eater would rather spend his life studying holy books than doing press-ups and deep-breathing exercises, or if he prefers to devote his life to things of the spirit rather than of the flesh, what can the biochemist say then? The pig nutritionist has things so much easier: he knows exactly what the ideal bacon pig looks like and what the perfect porker should weigh.

6 Examine the width!

Where King Midas went wrong was in developing a talent to turn *everything* he touched into gold. His story is an instructive one for us, in spite of its inherent improbabilities. How much more sensible it would have been if he had exchanged his general-purpose golden touch for something more selective. Had he done so, he could have amassed a reasonably sized fortune in his spare time and then been able to spend the rest of the day with his daughter.

For ourselves, in this generation of technological activity, science is the Midas touch with which we can turn rocks and stones into gold (particularly if the rocks and stones deep under the bottom of the North Sea show the way to oil-bearing strata). And we too, like the king in the fable, must guard against the danger that when we kiss our wives and fondle our children they will stiffen before our eyes. Once upon a time, people who strove to be gentlefolk possessed some idea of the kind of life they admired. These notions may have been wrong. The politeness, forbearance, sensitivity, education, self-sacrifice and internationally acceptable accomplishments would very likely be unpalatable to modern taste; but when the mother twittered urgently to her son, 'My dear, I don't mind what profession you take up as your life's work but, please, not trade!' misguided as she may have been in denigrating any honourable occupation, she was expressing a distinct point of view about the quality of the life she thought it appropriate for her son to aspire to. True, that at the time she said it, miners receiving a pittance for their work were labouring long hours in the pits and women working for even less were sweating in garment factories. Even for these, however, the target remained as a glittering star. 'One day,' said Sam Weller, 'I may be a gentleman me'self. With a garden of me' own and a pipe in me' mouth.'

Within the corruptible body of a man or woman, about whose substance, structure and biochemistry we know so much, there exists an immortal, immaterial, questing spirit. In just the same respect at the centre and core of science is an intellectual idea, the notion and understanding that gifted imaginative men have gained about the nature and substance of the universe. This understanding is doubly precious, first because it has been obtained by the imaginative ordering of the scattered confusing things that are to be seen, touched or otherwise apprehended; and secondly because the ordering which has been conceived in the head of some fallible person – the scientist, no less – has turned out to be a true representation of the way nature really does work. A man has to earn his living, and science as it is understood in the hard utilitarian world of today, in which the target of Man is economic prosperity rather than salvation, must also earn its keep. The pursuit of science is no longer an inexorable commitment. Unlike the painter, musician or poet who even at the risk of poverty and starvation cannot stop himself from making pictures, music or poetry, most scientists undertake science for a material end. But while the purpose for which the research stations are built, the radio-telescopes erected, the particle-accelerators constructed, and the mountains, deserts and oceans surveyed and studied is to establish a prosperous society and insure that the community's wealth shall continuously grow, there has in recent times been an awareness that there are questions to be asked. As a community we are sure that we need nuclear power to support industry and commerce or, if not nuclear power, something newer, cleverer, perhaps even something that science has not yet discovered. Yet, even in the enjoyment of the riches and comforts that science can bring, the antibiotics, the plastics and the pocket calculators, the questions bubble up. What, we are beginning increasingly to ask, has the effect of science and technology been on the quality of life? And what kind of life are we making for the years ahead?

Arising out of these another question comes to mind. To what extent is the ordinary citizen wise to place on the shoulders of scientists the responsibility for deciding whether to pursue or not those things that science makes possible? How to decide whether something that science makes possible will improve life? What kind of life will people find themselves living when it is done? This is a difficult, puzzling, fuzzy question because it demands an insight which no individual can hope to possess. For instance, even after more than 200 years' experience of the steam engine it is hard for us to decide, first of all, to what extent the steam engine was a scientific product; and, if it was, whether trains, on balance, made life better or worse. In the age of railways some great landlords and the mayors and corporations of some cities resisted the penetration of railways into their territories: today, in a period of motorways and airports, these same people are the most likely to exhibit a romantic attachment to dear old shiny steam locomotives. It is not easy to draw up a balance sheet of benefit and harm when we reflect on the plentiful supply of goods of all kinds available to almost every social group within the community provided by the factories which steam engines first made possible. And if the corporate wisdom of the nations looking back over two centuries cannot reach certain answers, how can we expect scientists to foresee what may happen when something untried and new is to be assessed.

Scientists by their very training and by the continued

practice of their art are required to give close undivided concentration to the detailed minutiae of their research: of all people they could be taken as the worst suited to judge the quality of life as ordinary people would assess it. This is not to say that there are no scientists who are deeply concerned about the effect of their discoveries on the social scene. For example, Professor D. J. Bernal, one of the greatest and most sensitive of British scientists, provided in 1969 an endowment to support a lecture to be given under the auspices of the Royal Society of London on some aspect of the social function of science. In 1976, this lecture was delivered by a distinguished Soviet scientist, Academician P. L. Kapitza. And so it came about that with all the pomp and solemnity of the ancient foundation, under the chairmanship of the President of the Royal Society, Dr Kapitza delivered a lecture on 'Scientific and Social Approaches for the Solution of Global Problems'. As could have been expected from so learned a man and so great a scientist, Kapitza reviewed the approaching energy crisis as supplies of coal, oil and gas inevitably diminish as the centuries pass. The solution he proposed with all the logic of his understanding of the laws of thermodynamics was the construction of more and larger nuclear power stations. He reviewed the impending depletion of sources of minerals, of iron and copper, aluminium and phosphors. He showed in charts and tables the pollution of the air, the sea and the land. 'It is evident,' he said, 'that, as the problems affect all nations, the solution to them is the concern of humanity as a whole.' And he concluded that 'the only efficient way to solve them is by a scientific approach'. 'The main difficulty which one meets,' quoth Kapitza, 'in attempting a solution is the social side of the problems which is closely connected with psychology and the structure of human society for which the scientific basis is still vague.' He then went on to outline his scientific solution. In the main this was a discussion of how best the growth of the human population might be controlled, not only by finding means to limit the freedom of young men and women to be fruitful and multiply, but also to arrange for citizens showing signs of wear to leave the world to their successors when it became obvious that their social usefulness was at an end.

Should we be prepared to accept this apparently clear-sighted approach as a formula for the quality of the life which scientific understanding should bring about? Perhaps there may be arguments – particularly at committee meetings of the League of Grandmothers – about the criteria to be used in assessing who is to be defined as 'socially useless' and when. After all, in pre-scientific ages, there were here and there a few people who felt the occasional twinge of conscience as one more old woman, selected by the tests of the times as being a witch and, by definition, socially undesirable, was done away with.

Although we have not so far agreed to the rational disposal of the old as a 'solution' to our current problems of establishing the good life, we now agree that it is rational to deplore the science-stimulated increase in human numbers. We accept the aid of science to block the conception of infants we no longer desire, and we apply skills of scientific medicine during intra-uterine life to dispose of offspring already conceived. There are historical precedents: Pharaoh, King Herod and the Spartans all destroyed new-born citizens, hoping thereby to enhance the quality of life for the rest of the nation (or so it seems).

Perhaps it is not a matter of science at all. Although it is true that the subtle combination of steroids from which the contraceptive pill is fabricated is a brilliant exercise in applied science, communities were controlling their own numbers by a variety of means long before modern science was in being. The ordeals which young men were compelled to undergo during processes of initiation as adult members of their communities were in effect devices for restricting the community; those who had not undergone the trials were not considered fitted to take their place as heads of households. In many societies girls also were required to submit to initiation ceremonies. And the Victorian father, sternly asking the young suitor whether he was in a position to maintain his daughter in the degree of luxury to which she was accustomed, was by no means the first to exact a 'bride price' from the family of a potential son-in-law.

But if we deliver the scientist from the responsibility of deciding whether infants shall be allowed to live and whether sick 'useless' old people shall be persuaded to die, should we perhaps blame him for disrupting family life and by so doing damaging the quality of life? The scientifically trained paediatricians naturally prefer to deliver women of their babies among the sanitary clinical conveniences of a purpose-built institution, but then the dramatic family event of a birth at home, with all the excitement of the midwife's arrival, the tramping up and down the stairs with kettles of hot water, the gossiping presence of grandmothers, aunts and neighbours no longer takes place. We have lost one of the dramas by which life's brightest quality was attained; how can a balance be drawn between that impoverishment and the one infant which could perhaps have been saved had its mother left the friendly confusions and inadequacy of home for the efficiency of scientific birth?

To turn to infant feeding, there is a problem for which a dogmatic answer is equally difficult to find.

Has the application of science, as exemplified in the development of artificial baby food, harmed the quality of life? One can study this from the viewpoint of the mother, no longer having to put her baby to the breast every two or three hours; of the baby, enabled to draw its early sustenance – a delicate balance of vitamins, minerals, fat, protein, sugar, salt and water – from a sanitary rubber teat rather than from a nipple of flesh and blood; and of the rest of the household, the other children, father, aunts, grandmothers, servants and lodgers. No longer can they all coalesce round the presence of the lady of the house: she is probably not at home but more gainfully employed out at work as a computer programmer for the Tax Inspectorate of the Inland Revenue.

Artificial infant-food mixtures owing their composition to scientific knowledge were first marketed in any quantity around the 1920s. It was recognized quite early that the milk from different mammals varies in its chemical composition:

notably, cow's milk contains about 3.5 per cent of protein whereas human milk contains only 1.5 per cent; by contrast, human milk contains about 6.8 per cent of sugar compared with only 5.0 per cent in the milk of a cow. Clearly something needed to be done to cow's milk, whether in dried or liquid form, if it was eventually to be used successfully to feed human infants. It is interesting to recall that in the pre-scientific age Mrs Beeton in one of the early editions of her cookery book advised that mothers unable to feed their own babies or obtain the service of a wet-nurse should try to bring them up on ass's milk. It was sound advice. It is now established that ass's milk contains about 1.7 per cent of protein and 6.9 per cent of sugar, remarkably close to the average values for human milk. If Mrs Beeton had only known it at the time, the frantic mother had another alternative ready to hand in the fields: hare's milk has been found by scientific analysis to contain 1.6 per cent of protein and 7.0 per cent of sugar!

The positive and negative aspects of bottle-feeding as part of the good life, can be argued chemically and nutritionally, sociologically, affectionately, economically and commercially. The chemical and nutritional aspect has a long history. Some of the first advertisements for seemingly perfect mixtures used to claim to 'build bonny babies': they were later found to be deficient in vitamin D. In very recent times even the great scientific engine of the State itself was found to have distributed, as National Dried Milk, a mixture calculated to be incompatible with total infant demands: salt had to be removed from it. The fact that at least two generations of citizens had grown to what appeared to be healthy maturity after having sucked it in infancy, was, it seemed, more a testimony to the toughness and durability of human young than a comment on the accuracy of scientific assessment. The chemical composition of dried-milk mixtures does not seem to have had much effect on the quality of life. If rickets appeared in the early days of their employment due to too little vitamin D and, if twenty or so years later, a condition known as ideopathic hypercalcaemia arose due to too much, it must be admitted that some of the breast-fed infants developed rickets as well and others got too little to eat altogether. Similarly, infants who suffer from defects in the formula of National Dried Milk – and there is precious little evidence that any of them did – could easily be paralleled by those who sickened and died at their mothers' breast.

The sociological argument is much more complex. All sorts of women have babies: intellectuals, artists, administrators and others who are incapable of administering anything. There are dominant masculine women and subservient feminine ones. There are maternal women who adore babies; there are others who dote on dogs. There are those who, if they do not actively dislike infants, at least find their company profoundly tedious. But surely in the light of this it can justly be argued that artificial baby food, the elegant product of food science, by extending the area of choice for all mothers without significantly affecting the nutritional status of the human infant, improves the quality of life to a significant degree.

And the economic argument – an argument frequently used to assess the wellbeing of a community and hence the quality of the life it leads – can it be taken to justify the bottle feeding of the community's young? The production of substantial tonnages of dried cows' milk and their compounding with a judicious mixture of appropriate ingredients; the packaging, labelling and marketing of the combination; the purchase of bottles, teats, measuring spoons and saucepans for the heating of bottles in the middle of the night must obviously involve a significant economic outlay. This must be related to the benefits of a prosperous baby-food industry and to the valuable transfer of a major labour force (as nursing mothers might, if they chose, become) from the nursery to the shop, the office, the factory or the ticket-collecting booths of the nation's underground railway stations.

How much easier it would have been either to say that bottle feeding did benefit the health of infants and must therefore be an improvement to the quality of life, or that it injured the infants and must therefore be bad. In truth, however, health (in its strictly medical sense) has little to do with the question. We are therefore left with the imponderable assessment of whether the baby suckled by its mother grows up into a happier and more valuable citizen, and whether a woman at home with her child enjoys a better life than one who is exerting her talents in the wider world.

Another unanswerable question often hotly debated is whether food science and technology, by which articles are frozen, dehydrated and canned, packaged, formulated, processed, stabilized with additives, thickened, emulsified, vitaminized, flavoured, coloured, even 'en-nobled' (to use a term invented by the late Professor Platt) improves the quality of life in a society or causes it to deteriorate. The first attack on foods sold in the supermarkets by those who feel that such foods damage life's finer qualities is that they do harm. 'Consider,' such people say, 'how many drugs and laxatives people consume today.' 'Consider also,' they go on, 'how healthy and nourishing food was in the past before science took a hand in its production.'

As with the baby-food debate, this argument has little evidence to support it. Adulteration and infection of food was more widespread and more dangerous in the days before science was available as a tool to analyse and separate items and measure the amount of water added to the milk by unscrupulous vendors; the amount of sand in the sugar; and worse abuses only brought to light in all their horror when scientific detection became possible in the 1860s. The refrigeration science makes possible today together with the understanding of microbiology has deprived three happy generations or more of ever having smelled a bad egg, stinking fish or decaying meat. At the same time typhoid, diphtheria and – within our own lifetimes – tuberculosis of the bone among children (from drinking infected milk), undulant fever and anthrax have retreated in importance as threats to the good life (or, to life itself) – to about the same degree of hazard as being struck by lightning.

A frightened generation prepared to be gleefully alarmed by being told in the morning paper that mono-sodium glutamate is poisoning the nation knows perfectly well: (a) that the Japanese have been eating it for centuries; (b) that we have too; (c) that the scary appearance of its name listed on that can of oxtail soup was put there solely at our own wish when we insisted that it should be so listed; (d) that 'the government' – meaning the long and honourable succession of Public Analysts who have successfully kept watch for a hundred years – would make their objection known should it really be poisonous; and (e) that to maintain the quality of life we desire we insist on the tastes, smells, consistencies, colours and sounds (who would accept silent celery?) being of the standard we desire. The very gentleman who so angrily complains about the tastelessness and lack of nourishment in 'miracle whip' or the like would be the first to complain about falling standards if he were not served with horseradish sauce to accompany his roast beef.

If science applied to food does damage the quality of life it is certainly not because it harms the health of those who eat what it helps so signally to provide. Brown-shelled free-range eggs improve the quality of life in its aesthetic rather than its nutritional dimension just as horseradish sauce and brandy – which of all liquors is amongst the richest in methyl alcohol – do. There are, however, two respects in which science applied to a community's good could be argued to affect, for good or ill, the quality of life. It can and does affect the nature and character of mealtimes. It also affects both the reality and the sentimental myth which people – and particularly people belonging to a mainly urban civilization – hold in their mind about farming. Meals are transfigured by the 'meal-in-a-bar', and farming becomes agribusiness.

A meal is a social experience. The child brought in from play, made to wash its hands, sat down in its place between a younger brother and grandmother, taught to eat with good manners, given the exquisite delight of its first mouthful of treacle sponge, its first taste of turkey with roast potatoes, stuffing and bread sauce, joining in the family conversation, revelling in the paper hats and crackers of a feast day is learning about its place in the community and storing up precious memories for the hard cold world outside. The young couple sharing together the entrancing privacy of a meal in each other's company, the regimental dinner, the wedding breakfast, the city dinner – all these exemplify the contribution of food to the good life. Science occupies a useful rôle or, at least, has a neutral influence in so far as it provides for supplies of vitaminized margarine; oranges, tomatoes and green peas at every season of the year; or a deep-freeze cabinet, a microwave oven and a supply of dehydrated milk and 'instant' coffee. When, however, 'convenience' foods are pushed to the extreme – which they are by the 'meal-in-a-bar' concept – in place of the chattering family, the school dining table, and the luncheon party of friends we have the suave stainless-steel fronts of a battery of vending machines; from them uniformly coloured, standardized articles, each on its cardboard plate or recessed into its press-shaped aluminium foil tray, can be obtained either hot or cold by the insertion of an appropriate sum of money – money! The damage to the quality of life is not done by dangerous additives or a lack of one vitamin or another. It happens when the consumer (and that is what he or she has become) eats the food in a hurry, standing up, driving a car, or watching television – alone.

The science of botany, of plant physiology, of genetics involving plants or animals, the chemistry of soil and the biochemistry of living cells, such science made potent by engineering and the whole applied to agriculture exerts a potent, subtle and pervasive influence on the quality of life. On the one hand it is dramatically beneficial. Life's quality can only become better when food is plentiful, diverse, pure and available all the year round, without the need for human drudgery. That this is so for technologically advanced communities is to a large measure due to scientific understanding of biology, chemistry and physics (through which the great machines that do the work become possible), much of which has only been won within the short span of living memory. At the same time the land – what is now fashionable to call 'the countryside' – possesses a peculiar, oddly deep, significance to the human spirit even when that spirit is housed within the person of a town dweller whose only contact with the country is acquired during the brief span of a package tour by bus. The rumpus of the 'enclosures' of the eighteenth and nineteenth centuries involved more than economics. People's feelings about the whole quality of their lives were involved. Now that once again the character of the country-side is being changed by the removal of hedges and boundaries whose erection was so bitterly opposed two centuries before, similar emotions are aroused.

Haymaking, harvesting and threshing were always technological acts. Once, however, they were social occasions as well. Farm workers were always paid, to be sure, but whole families were out in the fields once helping to bring the harvest home and at the same time blurring the split which today of all other things most woundingly damages the quality of life, namely the separation which people make between what they do grudgingly for money, categorized as work, and what they undertake with joy and enthusiasm for living, described as leisure. The migrant workers moved from place to place, living in untidy camps here and there as the seasons changed, picking grapes in one place, tomatoes in another; they enjoyed life of a higher quality than what came to be their lot in the shanty towns to which they flocked as a last resort when the work they did before came to be done by one of the new sophisticated products of technology. Elaborate potato harvesters, fitted with x-ray detectors to distinguish stones from potatoes, now perform in Scotland the task once carried out by a multitude of men, women and children (let out from school for 'potato holidays'). The raspberry fields no longer echo to the voices of people; instead there is a machine with a multitude of arms which shakes the fruit into a plastic chute. There are in the world only two dozen or so tomato harvesters, each one bigger than a diesel locomotive. There is no need for more; they are sufficient to harvest all the tomatoes in the world, as the lone tomato-harvester driver throbs backwards and forwards – in California, in Italy, or in Australia, it does not matter to him where he is, he belongs nowhere and works on contract – across huge ranches of tomatoes.

The beneficent revolution brought by science has been very rapid. Livestock, wheat fields, milking time, feeding the chickens, and taking the produce to market, all these have undergone abrupt metamorphosis. Inevitably now large, trans-continental vehicles are employed, that seem so unprepossessing to the citizen peering out through his parlour curtains. The quality of life of both food producer and of bystander from the town has changed. We look at the country through the wound-down window of our motorcar and find it different from what it used to be; but to what extent has it become worse and to what extent better? After all, the farmer *chooses* to be 'efficient' in order to be able to buy his nicely appointed house with all its desirable technological aids to modern living; just as the town dweller chooses to buy the motorcar from which he can look out at the growing corn.

There have been many philosophers and writers who have cogently argued that a city can provide a quality of life superior to anything to be found elsewhere. A Roman senator could enjoy a period of residence in his country estate but for the quintessence of life he must visit Rome; the medieval city states offered great benefits to their citizens in trade and commerce, culture, learning and society; the French could claim Paris to be the queen of cities; he who no longer loved London, stated Samuel Johnson, no longer loved life. Science, by which stronger concrete, plastic wall coverings, glass walls, air conditioning and prefabricated partitioning have become possible, enables buildings to be taller, bigger and less familiar to the older eye; it has changed the character of new cities and inevitably affected the quality of the life of those who live in them. In 1976, a bizarre discovery was made by two scientists, M. H. Bornstein, working in the Department of Psychology at Princeton in America and H. G. Bornstein, working in the Max-Planck Institut für Psychiatrie in Germany. These two carried out a study* in a number of cities in France, Germany, the United States, Czechoslovakia, Greece and India. The places they chose had populations ranging from fewer than a thousand to more than a million. In each place, a distance of 50 ft was measured along the pavement in 'func-

* *Nature*, 259, 557, 1976.

tionally parallel sites (in) "Downtown" or commercial areas like Wenceslas Square in Prague, Flatbush Avenue in Brooklyn, and Rehov Yerushalaim in Safed'. Then 309 people, walking in either direction 'alone and unencumbered' on a dry sunny day when the temperature was about 75°F (21.4°C) were in each place discreetly timed as they covered the distance. When the whole experiment was completed it was found that on average the bigger the town, the faster the people walked. In a town of 1,000 population the average speed was 1.8 m/hr; in a bigger place with a population of 5,000, the people walked at 2.2 m/hr; in a city of 100,000 the inhabitants walked at 2.9 m/hr; and in a metropolis of 1,000,000 the pace had quickened to 3.5 m/hr.

An ordinary person reading the Bornsteins' report must feel admiration for their extraordinary diligence and pertinacity at having collected all this statistical information about the walkers of Brno, Bastia and Brooklyn; Itea, Iraklion and Athens; Munich, Netanya and New Haven; Prague, Jerusalem and Dimona; Corte and Psychro; but what are we to deduce from the results? One talks about the 'pace of life'. Here in precise and literal terms we have a measure in miles per hour of the increase in that pace, as towns grow bigger. Is this good or bad? Should we welcome the brisk vitality in Prague and Jerusalem and deplore the strollers of Psychro and Itea, or ought one to envy the contemplative Itea inhabitant who gets about so much slower than the busy walkers of Brooklyn and the hurrying Munich pedestrians? What do the Bornsteins say? The results 'imply that demographic circumstances affect the quality of life, psychological experience, and motor action of humans' their report runs. It is good to find these conclusions supported by meticulous observations and measurement. Nevertheless, the question remains: is the quality of life better for places where walking is fast – or where it is slow?

The growth of urbanization has been one of the most astonishing features of the present century. The reason for the phenomenon is hard to discover. Could it be that the quality of life in cities is deemed to be better than that of life outside? Intellectuals who essay to study the problem may be loth to accept this answer because it is generally assumed by such as they that city life is disagreeable. Why then have increasing proportions of the people in almost all nations, freely and without compulsion – sometimes indeed against the wishes and appeals of their rulers – opted to live in cities? In Europe, London, Paris, Rome and the rest have grown inexorably bigger. Across the steppes and prairies of wide-flung spaces in the United States and Soviet Union alike the people have shifted into the towns. In South America too, as in Asia and Africa, men and women have voluntarily clustered in their millions into great metropolitan communities. It is said that life in New York, Tokyo and greater Birmingham is sordid and dreary for the poor: why then have they voluntarily moved there?

Money can in some respects be taken as the unit by which human desires can most succinctly be measured; hence what an economist discovers should have some bearing on what people consider desirable. Irving Hoch, an economist, attempted on a broader scale what the Bornstein's had done to find out how people were affected by bigger and bigger cities. Hoch has produced hard evidence* to show that as cities become bigger the value of land increases and rents rise. Both, we may argue, have become more desirable. Wages go up, so keenly do those who live in such places and run factories there desire new people to join them. Do the people who come into big towns and thus make them grow bigger do so because of the big wages they know they are going to earn? Do they also know that this increase in earning will be counterbalanced by the higher cost of food, housing and transport? Do they come just the same hoping to enjoy the diversity and variety of choice only available in a big city? These are too difficult questions for the economists to answer.

Science is the factor, the influence or power – call it what you will – that makes size possible. Whether or not 10 million people in Paris have a better quality of life than 1 million in Glasgow; whether 20 million (or whatever the figure is at the moment of writing) in Greater Tokyo are

* *Science*, 193, 856, 1976.

happier people than all those in Greater London, or São Paulo; it is true to say that none of these huge clusters would be possible without science and the technology it makes possible. There are the telephones and the water supply; the sewer pipes and the urban transport system; the juggernaut lorries to bring the food into the stores and the refrigeration equipment to keep it there; the radio sets for the people to listen to and the broadcasting stations to enable those who rule the people to tell them what to do. But did science start the process of aggregation or come as a result of it? Size affects the quality of life in many ways other than through the influence of bigger or smaller towns, whatever that influence may be. A family that lives in a small house, even though it may be (or may come to be) called a slum, establishes an emotional relationship with it – and the quality of life is more an emotional quality than a technological one. Less so do a family in a far more sanitary, warm, technologically advanced dwelling unit in a very large building. It is as socially neutral as an aluminium smelting plant – a construction with which no single person can build a human bond. In the dear old slum, a man could put up a garden shed; in a structure designed to rehouse that chilly modern entity, 'the community', he cannot even drive a nail through the polyvinylchloride wall cladding.

Many of the conveniences and advantages people enjoy in an age of science make large-scale cooperation essential. Even if it is decided that nuclear power stations are too dangerous and ugly in a community that values the quality of the life it leads, electricity is essential to its social welfare. To win oil for making the electricity from depths of the North Sea demands a great communal effort. In this very fact may lie the answer to the questions we have so far hesitated to answer in this chapter.

The unease felt by the citizen who spends his life in a great formless office block is due to his remoteness from the motive force that makes the system go and from the purpose it pretends to serve. To be part of a 'work force', directed by 'line management', itself obedient to 'middle management', which moves or does not move on the instructions of a 'subsidiary board', directed this way or that by a 'main board' – this is to feel one's very soul wither. The shrinking which people feel when confronted by size, by large, ownerless structures or by big impersonal organizations to which it is not possible to form human attachment – is due to the nature of the organization and not to the sophistication and subtlety of the science involved in their operation.

Science is, it seems to me, merely a servant of the community. When people say that science is affecting, for good or ill, the quality of the life of a community, they really mean that the community is not using science wisely. In 1848 *The Times* of London vehemently opposed the proposal to lay the city's first main sewer, appearing to feel that the freedom to die in one's own way without civic interference was one of the elementary rights of an Englishman. 'We prefer to take our chance with cholera and the rest than be bullied into health. England wants to be clean but not to be cleaned by Chadwick.'* In 1861 the Prince Consort died of typhoid as a result of this bizarre preference. If the community does not like air pollution it can, if it chooses, cut off its own comforts and wealth by an attack on science and technology and the closing of factories, or it can after due thought introduce a Clean Air Act. If it sees fit, a community worried about numbers can introduce with appropriate scientific assistance contraception or abortion. Surely this is better than the ancient primitive remedy of infanticide?

The judgements people make about quality of life are as much to do with what they feel it to be as with what it is. Two people *can* be happier in a cottage than in a palace. If the employees of a multi-national company, well paid though they may be, feel themselves drained of vitality, it is spirit rather than substance that has damaged the quality of life they feel they lead. To win oil from the North Sea was hard, dirty and dangerous, requiring the development of scientific understanding and engineering skill of an unusually high order. It could only be attempted by large organizations. Yet because of the rigours of the undertaking to be engaged in the

* Magee, B., *Towards Two Thousand*. Macdonald, London, 1965.

enterprise was an enrichment of life for the enterprising men who undertook the task and (when they allowed it to do so) lifted the spirits of those who witnessed such a bold, dangerous and magnificent operation. Who could ever have conceived the erection on the bottom of the sea of great pyramids three times as lofty as the Eiffel Tower?

7 Poverty in style

The definition of poverty is more subtle than one might at first imagine. Only if a man's resources even in good times are so scanty that he is compelled to spend his days grubbing for a bare subsistence of a handful of food from an inadequate patch of impoverished soil is poverty necessarily synonymous with starvation, homelessness and rags. There are today considerable numbers of irritated and frustrated welfare workers: they went out into the far corners of the earth, filled with lovingkindness, vitamin tablets, tents, blankets and insecticides fully prepared to tell communities less fortunate than their own how to run their affairs, manage their farms and feed their children – only to find that the impoverished people they were so anxious to serve had their own ideas about what they wanted. Sometimes even when they were palpably underfed, the items they held in highest esteem were weapons with which more effectively to quarrel with their neighbours – or whisky.

There are poor people in Great Britain and, for that matter, in the even greater United States. In the main, however, they are not naked nor are they unfed. In such countries poverty has a different meaning. Such people can legitimately describe themselves as poor when they are unable to afford a holiday or a television set. Even though we may initially feel shocked at it, there is no escaping the sociologists' definition that poverty is a condition in which people are unable to afford to maintain themselves at the standard of living they expect – that they feel they ought to be able to enjoy at the level of society to which they feel they belong. There is a significant difference between a distressed gentlewoman and an abandoned working-class wife, as there is between a ruined company director and an out-of-work journeyman; but all four would consider themselves to be suffering poverty. Once upon a time poor girls had no shoes; today they are poor indeed if they are unable to afford to buy themselves nylon tights of a sheen, colour, texture, elasticity, durability and design that would have been the envy of every society hostess of the 1930s. Remembering what American soldiers serving in Europe were able to obtain in barter for a pair of nylon stockings makes one wonder what riches these would have commanded in the pre-scientific centuries.

If we accept that poverty possesses no absolute meaning but is an amalgam of what we have and what we consider we need, it is obvious that there are two ways to raise our standard of living and level of wealth: the first, upon which people and governments alike have concentrated so much thought and effort, is to acquire more things – that is to say to continue to increase our wealth; the other is to reduce our expectations. In just the same way that science has been enormously successful in increasing our wealth, it can, if we only apply it in the right way, be equally potent in enabling us to reduce our expectations without at the same time having to change our way of life. But in fact the process has started already. The very potency of science and its effectiveness in making us rich or, as we have become accustomed to say, raising our living standards, increasing our productivity, and expanding our areas of choice by the provision of new, sophisticated luxuries – all this may of itself be causing us of our own free will to restrict our expectations.

When we were poor we had, perforce, to walk. We touched our hats as the haughty nobles drove by in their carriages: 'If only we too were rich,' we said, wiping off our clothes the mud splashed up by the carriage wheels, 'we too would drive.' And the time came. Factories were built; wealth spread throughout the community; we all took our places in the complex industrial society. First came the steam engine, later all the electric motors running the machine tools we had learned to operate. We bought a bicycle. This machine was itself a miracle of drawn steel tubes, ball-bearings and sprocket wheels cut on a device representing something quite remarkable in applied geometrical mathematics. It opened up the countryside which without this invention might have fallen into irrevocable decay after the coming of the railways.

Bicycles had their period of flowering followed by an autumn of specialized use. Then came the motorcar. Scientific fertility caused the automobile to grow and flourish at every point. The efficiency, power and reliability of the internal combustion engine is a testimonial to engineering skill and design. It is certain that the drive and intelligence of brilliant minds will stimulate further advance; in twenty years the machine of today will be a quaint example of industrial archaeology. The improvement in tyres, had it happened alone, would have been held up as the principal glory of an intellectual age. In the multiplicity of discoveries and their instant application we have taken what the tyre scientists have done as if it were nothing.

Universal prosperity and the effectiveness of technology, together with the remarkable success of social engineering eventually permitted most people in the community to possess a motorcar. The sheer multiplicity of cars began to destroy their usefulness and the pleasure they provided. Already the rich started to discover an enhanced contentment and peace of mind in *not* possessing a vehicle of their own. Observing this, the more perceptive members of the poor discovered that there was little disadvantage and certain solid benefits in staying still most of the time and, when a journey became necessary, making an event out of it. Bicycles started enjoying an Indian summer. The habit of walking is on the increase.

One of the marvels of applied electronics is broadcasting. No one who did not live through the upsurge in true wealth which broadcasting brought to life can realize quite what it was like. The wireless! The word itself illustrates the magic of what took place. The telegraph, soon followed by the telephone, were curious, strange advances themselves; but the conveyance of electric pulses along a wire was something which even a non-scientist educated in the classics could comprehend. To send pulses through the invisible and impalpable ether seemed to be a new and remarkable achievement difficult for the poor human intellect to encompass. The crystal set, curious, simple, yet subtle, was a revelation. To crouch with a pair of earphones pressed tight over one's ears; to hear, faint yet clear, when the 'cat's whisker' – a tiny twist of fine wire – touched the right sensitive spot on the crystal, that distant yet friendly voice saying 'This is 2LO calling'; this was indeed to enter a new era. In a little while there were valves and a 'loud speaker' – a small black horn standing on its metal pedestal. The process of scientific sophistication had begun. Radio had come to stay.

Then came the transistor. The laborious need to carry heavy accumulators to a nearby garage for recharging was no more necessary. Freed of electric flexes and the weight of batteries radio sets could be carried anywhere. The empty air was indeed filled with voices – and with music too. Today the pop music ceaselessly dribbles all day long. But interesting things have happened. Receivers through which this music from the air becomes audible have become available virtually as of right, like a vote or a National Insurance Number, to all. Side by side with the cheap radio receiver has been the expensive complexity of the various and prestigious pieces of 'hi-fi' equipment, the evolving toy of the still rich. At the same time, steadily growing as time passes, a new category of people has appeared, who sing and play the guitar and in other ways enrich their lives at little cost. The new society is diminishing the need for the cohorts of people who produce music in the broadcasting stations, man the reception rooms and the security posts, service the land lines and the towers from which the radio waves are diffused – all by singing and playing themselves.

It is not long since the fashionable sin of people who opted to reduce their expectations rather than increase their wealth as a means of meeting the strains of technological society was 'absenteeism', and in its day it was considered to be a heinous sin. A coal miner might find that he could live comfortably on three-quarters of the sum he was offered, and that to take the whole would embarrass him; corrupt his wife with a superfluity of vacuum cleaners and flights of artistic china ducks; and cause envy to rise up in the otherwise blameless bosom of his brother who had gone into the church – but if he decided to take a day off to sniff the air and prune his roses he was accused of absenteeism. Worse, he was found guilty of being anti-social. And to sin against society is, in the age when the dogma of science rules, of a different order of culpability and abandon than to sin against God – an expression, indeed, rendered virtually empty in an 'enlightened' community.

I do not wish to exaggerate. As I write, this aspect of the trend towards poverty in style, although perceptible, is nevertheless a minority taste. In the main, the application of scientific ideas to increase an individual's wealth is pursued regardless of the strains of doing so: many are the men and women who, in order to share in such wealth, are prepared to sink themselves into the composite amalgam of the 'work force'; few are those who, having earned enough, give up the rest and go fishing, or paint pictures, or even take a week off to celebrate midsummer. It is not so in New Zealand. There a segment of the population have the unusual notion that money is for spending. Although perfectly capable of operating electronics factories and petrochemical plants they still think it sensible to take a fortnight's leave of absence – not only without pay, but in the teeth of hostile admonitions by the manager – to enjoy the funeral of a great uncle in the village where they were born. But then the New Zealand citizens who do this are Maoris.

The voluntary move towards poverty in style is more clearly seen in group actions by bodies of people than in the changed attitudes of individuals. This is to be expected from my definition. To impoverish oneself alone in pursuit of greater happiness is to feel poor indeed – poverty being the inability to keep abreast of one's own particular Joneses: but if the whole group decide that riches do not bring happiness, the ranks of the Joneses remain in line. A modern technological community cannot function without a sufficient supply of electricity, whether or not it is produced from coal, oil, water, nuclear energy or – should this ever be possible – the waves of the sea, the rays of the sun or the wind. For the people to be rich, the electricity, as viewed from the standpoint of their level of wealthiness, must be cheap. Yet today the citizens of almost all technologically advanced communities are prepared, if not to accept, at least to consider an increase in the cost of their electricity without there being (or even the consideration of how there might be) any corresponding increase in the people's wealth.

Science decrees that to make cheap electricity the power stations where it is produced should be few and big. There must therefore be high-voltage cables to bring the power to the people. But the people are beginning to say that they do not like the look of the cables. They are almost beginning to say (even though some of them do not quite know what they are saying) that they are prepared to impoverish themselves to the extent of paying about thirty times as much per kilometre for underground cables carrying the electricity at a lower voltage simply because if they went into the countryside they would not want to see cables hanging over it. But perhaps it may be argued that this is not an altogether clear example of people expressing a conscious desire to be poorer.

A South African engineer, W. W. Campbell, diagnosed in 1973 a condition which he called 'the UVT factor' (us-versus-them) from which he deduced that there might be those who felt that power stations built themselves and that what the complainers wanted was to go on becoming richer while enjoying the satisfaction of doing what would make them poorer.

Whether all those caught up in the new tide of feeling understand its significance or not, there is little doubt that what is happening represents a change in the flow of history of the technological age. In the United States, the country which of all others has embraced the use of science in human affairs, the new direction is most striking. Although wealth and money are so strikingly set up as the goal of human endeavour – whether in business, in golf, in the practice of medicine or in politics – one finds the nation passing laws requiring a report on the 'environmental impact' of almost any proposed innovation from the building of a new power station to the making of plastic tiles. And this report is required to be so detailed and elaborate and to be submitted to so much close scrutiny so often repeated that the cost of initiating anything new becomes too high; the upshot of each submission is so unpredictable that in many instances the new project is not started at all. We see communities of science-orientated people who voluntarily choose life without a factory where there used to be fields, a refinery where fishing boats once came in from the sea to anchor, and without dipping wires stretched above the valley where no wires stretched before.

In many respects what we see happening is not new. The people who built cathedrals and palaces did not always do so of their own free will, yet they were quite well aware that the effort and treasure required to accomplish such expensive operations would never be repaid in a material sense in money. Although one might say that the building of ornate palaces and great churches impoverished the people who built them, their poverty could truly be described as poverty in style.

It would be an interesting exercise to try to assess the

cost of the efforts expended by environmental-preservation societies of various sorts; the value of the land tied up in expensively policed wilderness; the value of trees not properly cropped by foresters and timber merchants but allowed to grow old picturesquely until they fell down – all this investment dedicated to persuading a dozen pairs of ospreys to nest in the British Isles. The preservation of these ospreys is only partly excused as satisfying the aesthetic taste of the few people prepared to travel to the Scottish Highlands, haul themselves up into the specially prepared vantage points and look through the specially aimed field-glasses. It is also – and more weightily – justified on the grounds that the presence of the birds serves as a contribution to science.

This could be argued to be a happy reversion to the early nineteenth century when the great banks, whose purpose was without question to acquire wealth for themselves while caring for that of others, built enormous ornate classical structures to house their head-office staff. The cost of these buildings – many of which were very fine – could not be described as impoverishing the banks, but it did, even if to a small degree, make them less wealthy. Only within the twentieth century has the pursuit of wealth, commonly called the establishment of a healthy economy and a high standard of living, been tacitly accepted as the primary goal of society, a target towards which science and the technology derived therefrom should be aimed. The value attached to the non-mercenary is expensive to attain and consequently impoverishing, but concern for the countryside, the air, the waters of lakes, rivers and seas, birds and beasts – all this can be deemed an improvement on the single-minded pursuit of wealth. Since science and its deployment in the manufacture of transistors, plastics, computers and non-drip paint has been intimately involved in the mercenary activities of modern communities, the changed outlook in favour of a measure of poverty is equally a change in the public attitude to science.

Science is partly in favour of poverty when, at considerable expense, factory owners are compelled to fix scrubbers in their factory chimneys to limit the discharge of dust and fumes into the atmosphere. The same applies when they need to devote major sums of money to effluent purification: now they are starting to remove noxious substances from the liquid outflow, rather than discharging them all into the sea. Again, the cost of goods must inevitably rise, the affluence of those who need the goods consequently decrease and their poverty becomes greater, if money has to be spent on the installation of guards on machinery and of air extraction devices over grinding machines. The general acceptance that such things are good to do is comparatively new in historical terms although it now goes back three or four generations. Science is intimately involved here: it shows the existence of things which people did not know existed before, but which they are prepared to impoverish themselves to avoid. For example, asbestos had been used for a variety of purposes ranging from roofing to mats to protect the kitchen table from hot kettles. Yet when scientific study implied that certain types of fibres could be harmful to health – very different from the identification of a particular person whose health has in fact been injured by an asbestos roofing tile or a kitchen mat – men and women willingly accepted some impoverishment to abandon asbestos. Nay more. They urged the State to insist that everybody should similarly be impoverished.

Science is equally closely involved with poverty and its amelioration or aggravation in the wider arena of the civilizing of technological societies. Much of what it is customary to describe as 'welfare' has a scientific basis. If criminals of various grades were summarily executed the country would be enriched, since the drain on police guards, prison buildings, catering, entertainments and special electronic detection equipment would be avoided. That a solution to the problem of this nature is not even considered is in some measure due to scientific studies of deviant behaviour and the psychological influences experienced early in life, and to a statistical examination of the effect of condign punishment on crime rate. Hence – partly because of a general increase in respect for the rights of citizens, even deviant ones – the substantial expense of modern penology is accepted. Nevertheless, the national wealth is impoverished thereby.

There is a series of other charges which science, as understood by those in charge of public policy, indicates should be borne by the community. Houses must be warmed to a temperature assessed by science to be desirable; they must enclose a specific volume of breathable air; they must be equipped with water of a quality and purity also established on scientific criteria. Again, motorways must be smooth and wide, bisected by a crash barrier of scientifically accurate curvature and strength. And so on. We can gradually reassess each one of the scientific standards which advanced societies have set up for their own comfort, security, health and feelings of legal propriety. The reassessment is needed when the cost of all these standards, justifiable no doubt to a wealthy community, begins to impoverish people to a degree they cannot bear in times of hardship when their affluence has dwindled away. Left to themselves the individual men and women who make up society exhibit surprising powers of survival when money is short and goods are dear. Reassessment in a technological age is useful to them in showing them new choices of which they might not otherwise have been aware. It is most valuable, however, to those who make the rules and set the standards.

Many people even in a technological society are unlikely to be able to gather together enough money to buy their own house. Some years ago we decided that all the houses built by the community for the use of such people should be of a standard they could expect to enjoy, but it should not be too difficult in times of stringency to persuade these 'paupers' in our wealthy society to accept just a little bit of poverty. Community standards, for instance, could be set to insist that, in the cause of comfort, economy and food safety, every house to be provided for a citizen, rich or poor, *must* be fitted with that most elegant example of applied thermodynamics, a refrigerator. When poverty struck, the fridge might be omitted without undue fall in community happiness, provided *everybody's* refrigerator were withdrawn. The poverty of *not* owning a refrigerator would be understandably painful, but acceptable. Underfloor heating is another splendid example of applied science (the Romans achieved it at the time of Christ but only for the rich); the cost of operating it sometimes conflicts with the expense of other goods which people hold to be more desirable. So those who have it may not always use it.

The balance between poverty in style and the sort of poverty which people struggle to avoid is a delicate one. If the poverty of no telephone and no double glazing is perfectly acceptable, that of no bathroom and no electricity is very much less so. Public officials seeking to lower community expectations so as to free people from the toilsome burden of increasing social wealth must recognize that they may find a frontier that people will not cross. Nevertheless a wide range of options remains.

We can make public decisions to be poorer (in money terms) rather than richer in many ways: by choosing *not* to build factories and oil refineries in order to safeguard the continued existence of totally unproductive Golden Eagles and rare mosses; or by patching up the City of Venice rather than tearing the whole place down and building model flats for workers. But there are also quasi-private and private actions by which individuals can either opt for a degree of poverty or accept poverty in a stylish way. Here is an example of a quasi-public decision. We consider it seemly to pay ticket collectors on the London underground transport system an adequate wage sufficient to enable them to enjoy – even if at a modest level – the affluent society. Then we find that it is not possible to spare the money to pay as many of them as would be needed to sell and clip the tickets of all the people who desire to travel. So the community, making use of its considerable talents in electronics and engineering, install highly sophisticated machines to take the money and issue the tickets and other machines to examine them, open the gates and let the passengers through to the escalators – the lifts and their liftmen having long disappeared. And the passengers thinking nothing of the menial task that they would scorn as an occupation, happily act as their own ticket sellers and purchasers and, furthermore, do these things without pay.

Jane Austen at the beginning of the nineteenth century, Dickens towards the middle of it and H. G. Wells at the beginning of the twentieth have in their books much to say of the humble people employed at low pay to work in shops and, with all due servility, attend to the wants of customers. Most of these shop-workers have now succeeded to higher things. All of them have joined the rest of the community as customers who now themselves hunt around the shelves and put together the orders. As a child I often accompanied my mother out shopping, to hear at the end of the series of transactions the obsequious request, 'Shall we send it, madam?' Surely a voice from another age. Today, we content ourselves to do the work of the butcher's boy (if we can afford meat at all), the baker's boy and the vanman from the haberdashers, we hump our carrier bags home – and do so again without pay.

There was a time when, coming early to a restaurant, one could find oneself and the other diners outnumbered by waiters. There was always a head waiter to bow you into your place, and then with all due dignity – after an apprentice waiter had taken away your coat and umbrella – call up the real waiter who was actually going to do the waiting. He in his turn was assisted and supported by several more – one to serve the potatoes, another the greens, a third to push round a joint kept hot above its flickering spirit lamp. All these have gone except in those establishments to which one is driven, not by hunger, but by the occasional need for expensive entertainment. Ordinary people find it no hardship to become accustomed to the life of new-style poverty, and are quite happy to play their part in a cafeteria, themselves doing the work (of course unpaid) once undertaken by all those waiters. Science and technology inconspicuously support their endeavours. The hot dishes nowadays are kept hot by infra-red lamps and the cold ones kept cold by discreetly applied refrigeration while many of the dishes may well have been prepared months before when food supplies were plentiful, the cooked items subjected to the 'cook-freeze' process and filed away at −20°C until they were needed.

There has undoubtedly been a great social awakening which has not yet been fully heralded: there are now no tasks too menial; all may be undertaken by those enjoying stylish poverty without affecting the happiness of life. My parents believed that they could only live tolerably with the support of a hierarchy of subordinate servants. Even Mr Pooter, the hero of the Grossmiths' *Diary of a Nobody*, himself a clerk, believed that he had to have a 'slavey' to clean his house, do the cooking and, in the quaint phrase of the times, 'answer the bell'. Today, dukes, like the rest of us, think nothing of hauling the dustbin out of the front door and down the garden path onto the pavement. And none of us expect to be paid for being our own dustmen.

When paint was manufactured from traditional recipes and liquefied with linseed oil it took ages to dry, dribbled up the handle of the brush and needed to be applied three coats thick; a professional was needed to apply it. Today non-drip paint, making use of the scientific principle of thixotrophy does not make the brush into a mess and one coat covers almost any background. Thus the poverty of doing one's own painting and decorating for nothing becomes easy to bear.

The growing awareness that we can without discomfort *do* things in our role as happy and stylish members of the new poor is reflected in our dress. We find the population of the industrialized West increasingly wearing jeans. Partly this is a vague rejection of the price they would otherwise have to pay even for nylon, terylene and boots and shoes stuck together by adhesives rather than cobbled together by bootmakers; partly it is admiration of the romantic and efficient East. The result is we wear what is not so altogether different from those suits of dark-blue denim pyjamas which are the usual dress in the People's Republic of China, the most numerous nation on earth. It is interesting to reflect how poverty in style can be achieved either by the adroit application of science – as by the development of low-cost textiles, automated processes to take the place of people, and substitute foods – or by the acceptance of customs which are the common usage of communities which have always lived at a lower economic level and have known no better.

Long ago, the head nurse, the undernurse and the wet-nurse disappeared and the newly impoverished ladies did such nursing of their young as needed to be done for themselves, organized cooperative baby-sitting and – failing the ability or the desire to produce their own milk – bought something elaborated to be just as good from the chemist. The nursery too disappeared and in doing so brought about an economy in lighting, heating and rates. The elaborate perambulator of the Victorians and Edwardians was replaced by a simple collapsible device similar to a golfbag with wheels. In a short while this too will become a thing of the past: advanced thinking in paediatrics and the pinch of poverty have led to the introduction of a simple sack in which babies can be carried everywhere on their mothers' backs. We see the same device among the peasant communities of Malaysia.

Ivan Illich, in one of his more provocative moods, suggested that children were invented only a hundred or so years ago solely for the purpose of providing jobs for teachers. Be that as it may, the professionalization of teaching as an expensive arm of government propaganda to ensure the production of generations of acceptable citizens is already proving unduly costly. Parents are already finding the education provided by private (quaintly called 'public' by the English) schools too dear to buy; we can foresee the time when we will consider the provision of community education, despite forced economies on teachers, too expensive as well. When that time comes many parents may choose to undertake the instruction of their children themselves, and some may even discover that such intercourse with their children, even though it is imposed by the new poverty, is agreeable as well.

Gradually knowledge is spreading that the human thigh, as a machine for the development of mechanical power, is more efficient than many of the combustion engines in use today, but will it lead to any general replacement of the ever-more-expensive motorcars and railway trains by bicycles? Possibly a useful and lucrative development of the near future would be modern and more efficient versions of those vehicles of our ancestors, the tandem-bicycle built for two and the dramatic, powerful but difficult bicycle for four; this might take the form of a vehicle looking like a motorcar but escaping the burden of motor tax and the continuously increasing cost of petrol by recoursing to thigh-power. Accustomed as we are to the effete indolence of being propelled by the power of petrol, it may seem incongruous that in the future both we ourselves and the passengers in the back seats of our vehicles may expect to be called on to pedal the carriage along. Yet not so far back in historical times the passengers in a carriage expected to get out and walk to help the horses drag their load uphill. Indeed what has become no more than a saying was once the literal truth: they would put their shoulders to the wheel.

While science and the technical devices it renders possible provide comforts, conveniences and luxuries of every sort, the human happiness such things provide may be of doubtful value or may not be real at all. For the most part – just as beauty in the eye of the beholder – it may be evanescent in the mind of the possessor. The money needed to buy more such things, or even to keep up the supply of those which have come to be thought necessary, will eventually run out. The community's efforts to increase wealth enough to pay for all it believes it wants will prove to be beyond its powers. And at that precise moment the alternative of reducing expectations may prove to be more attractive and less revolutionary than it seems.

There are, of course, barriers to be climbed – some material, some psychological. I remember a fearful postman, anxious about Great Britain's food, whom I advised to eat more Yorkshire pudding (which he and the nation could afford) and less roast beef (which it seems they could not). He replied, 'But I *like* roast beef.' His solution can be provided by food technology capable of making artificial beef out of beans. This is an example of a material solution, in the fashionable jargon described as a 'technological fix'. A psychological solution would be a rise in the social acceptance of sausage or of some other mixture, whether new or traditional, incorporating low-cost food. There have in history been remarkable swings in the popularity and prestige attached to various articles of food. Winkles (although they have become increasingly expensive) have retained their low social status as a food for the working classes; oysters, as Dickens pointed out so forcibly in the *Pickwick Papers* (published in the first half of the nineteenth century), were once held in low esteem and were consumed primarily by the lower social echelons, but in the course of time they have come to be esteemed by the wealthy.

The history of bread shows how emotional rather than material are the motives which influence 'economic' behaviour. Two or more centuries ago, the softness and delicacy of white bread made it the choice of those who could afford to buy it. Poverty compelled poorer people to put up with coarser bread of rougher and more uneven texture. When in the nineteenth century advances in milling technology made it possible to separate the white parts of milled wheat from the darker fractions, comprising the husk and the fat-filled and readily rancid embryo, the poor also were able to treat themselves to white bread. Soon after, the educated upper middle classes, as gullible as any other type of citizen, got it into their heads that they needed more vitamin B and could obtain it from brown bread, despite the fact that of all social groups theirs was the most completely free from beri-beri, the disease of vitamin-B deficiency. Or they decided that they needed more roughage to move their bowels. Brown bread consequently acquired a social cachet. By such a roundabout way, should poverty strike, the minds of industrialized citizens will already be half prepared to accept the coarse, cheap bread from which their predecessors strove so hard to escape.

8 The bicycle culture

A man of culture is a remarkable creature. There he is, floating in a magic fluid which wraps him round and bears him up. Its iridescent pattern colours his feelings, beliefs and behaviour. Like a cloud in the sky, culture has shape and form, it provides glimpses of a clear horizon and a blue heaven in some directions and happy obscurity in another. But while it possesses some stability of form it is also subject to capricious and inexplicable changes. Not long ago, men of culture saw the world coloured by the classical books of Greece and Rome, by a proper study of mathematics and music, architecture and the furniture of their age. The man of taste, the privileged man, was contrasted with the vulgar artisan and with savages and barbarians. When, in Jane Austen's *Emma*, the vulgar Mrs Elton suggests that Mr Knightly should arrange a simple picnic with a table – a table, mark you – spread in the shade, he considers a cloth laid in the dining room more natural. 'The nature and simplicity of gentlemen and ladies, with their servants and furniture,' as he puts it, 'is best observed by meals within doors.'

Today the man of culture, although we are too timid to call him such, expects to find electric light and a flushing lavatory as part of the atmosphere from which he looks out on the world. This world also is a world of clocks, not merely the pendulum clocks of our grandfathers or even the chronometer of John Harrison by which navigators could find their way round the globe, but uniform clocks and watches, all beating as one everywhere over the earth's surface, checked each day by the implacable pips. A prophet can be telling the secret of right and wrong, happiness and sorrow or the coming of a new Messiah. Whatever it might be, the pips are paramount. The smooth unconcerned interviewer will cut him off in mid-sentence. 'Very briefly,' he will say, 'very briefly, Dr Einstein, what was your reaction when you first knew that you had discovered the Holy Grail?' And before the seer can get out another word the interviewer will be announcing the football results.

Our culture is a mist through which we peer out at reality; and this mist is coloured by particles of science. Science allows us to be quite sure that we know that all the stuffs on earth are the elements of chemistry and that the stars are composed of the same chemical elements too. Did not the spectroscope of Bunsen and Kirchhoff tell us so? And Newton showed the men of our culture that the motion and interaction of the parts of the universe we live in are comprehensible, nay explicable, to their rational intelligence.

The civilized scene, the comfortable cocoon of the technological world in which the man of culture lives is attractive and agreeable. He and his companions, whether they take their enjoyment at the football match and in the bingo hall, or in the library and in the Festival Hall, expect the telephones to go on working and the daily newspaper to arrive. A family car is an equally necessary adjunct to the good life. People do not expect to have to trouble themselves about the details of the processes by which a transistorized radio or a washing-machine and spindryer are made and how they work. They merely expect them to be there. They are vaguely aware that someone initially has to discover the scientific principles upon which these gadgets are based, and they perhaps admit that such discoverers are quite clever in their way; but if they happened to meet the men who did it they would not expect them to be particularly out of the ordinary. Nor would they be disappointed. The ordinary citizen is quite correct in his assumption that the people who work in the factories where radios are made, or even where Concorde aircraft – the very last word in engineering genius – are put together are not particularly gifted or unusual. Indeed, should a transistor-radio or a Concorde factory be put up in their part of the country they would apply for a job without any fear of the work being beyond their intellectual capacities.

Some time ago I took a bicycle as a simple example of a cultural object of today and allowed my mind to wander idly over a speculation. Suppose a new Adam, naked and alone, freshly created – as intelligent as any twentieth-century citizen but alone and unsupported by the complex impedimenta of culture by which we are surrounded. What would he have to do if he sat down in an empty England or France, Italy, Japan or Detroit, to make all by himself a three-speed bicycle with rubber tyres and dynamo-operated lamps from its raw materials.

Of course, by himself the poor chap could not do it. On his own, without the culture of his technological tribe all around him he would never be able to smelt iron ore, make the iron into steel and draw it out into tubes. Even in the present age, there are comparatively few of the civilized nations of the twentieth century who can make steel tubes. Aluminium, which was a rare scientific curiosity at the time of the Great Exhibition of 1851, would completely defeat him unless he made himself an electric furnace. Yet it is a matter-of-fact commodity to the technological culture which inherited the marvels that so entranced the Prince Consort;

his effigy still looks out over London from the plinth of the Albert Memorial with, as is only right and proper, a bronze model of the catalogue of the Great Exhibition in his hand.

Returning to his unattainable bicycle, the new-born, culture-free Adam would attempt in vain to fabricate his tyres even if the rubber plants in Kew Botanical Gardens were at his disposal; nor could he produce either the dynamo or the light bulbs it was to illuminate, any more than can half the newly liberated nations of the United Nations who even today come from a different and non-technological culture. But while the single Adam would have to go without his bicycle, no matter how intellectual and capable he might be, a crowd of very ordinary, not particularly intelligent, beer-drinking Englishmen have no difficulty making bicycles. And bushmen in Central Africa make quite sophisticated and highly stylized thatched huts. It is merely a matter of culture.

It is easy to make bicycles, radio sets, biros, pocket calculators, plastic carrier bags and all the paraphernalia of useful ephemeral objects – such as this year's model of a motorcar – that we take so unthinkingly for granted; easy because, while most of the children for whose public education we pay by our taxes pick up from it no more knowledge than they need to read the sporting papers and work out the odds for the 3.30, some of them learn chemistry and physics; and quite a lot more later on learn about machines and how to handle them during their apprenticeship. At the technical colleges, largely unrecognized and almost totally unpraised by the intellectuals who write the clever articles in the better-class newspapers, are the few who enrich the tribal understanding of every well-run industrial community. Fewest of all are the rare graduates and ever rarer PhDs at universities – the veritable medicine-men of the tribe – who strive to understand the mechanism of the universe. These are seeking out the quintessential Ju-Ju, the gods of the sticks and stones, the thunder and lightning, the sun, moon and stars, that make the machines run, the electric typewriters type and transmit the voices of the panel-game compères across the void. It is these rare few who must be understood by the community as a whole if the forces they command are to be harnessed to the benevolent production of deodorants and aspirin, television sets and semi-moist pet food rather than fifty-storey tower blocks by which the sky is darkened, foaming detergents smothering the rivers and streams, and the shattering roar of everlasting jet-propelled aircraft obliterating the God-given cadences of human speech.

It can be argued that science is science wherever one may be. This is, of course, true. Water is always H_2O, whether it runs from a tap in the Massachusetts Institute of Technology or trickles down the steep flanks of a holy mountain in Tibet. Equally can it be asserted that it boils at 100°C provided the boiling is done at sea level, regardless of whether the sea is the cool, rational Baltic, the drilling-platform-dotted North Sea or the romantic spice-scented ocean of the Far East. Move a coil of wire through the lines of magnetic force and an electric current will flow whether the experiment is carried out by a devout scholar in Pekin, a missionary in Entebbe (if there should be any missionaries left in Entebbe), or in a school classroom in Bradford. These things happen because scientific facts and what the textbooks say about them are true, are they not? And they are equally true whether the pupils to whom they are imparted are English, French or Iranian. The wealth of a modern Britain or a modern West Germany depends primarily on industry; such modern industry and the technology it uses depend to a significant measure on science; the ability to make use of science depends on there being enough people who have learned at school what science is about; does this mean that communities possessing a different tribal culture, perhaps one not using motorcars and aspirin tablets as avenues to corporate happiness, which want to change to our culture can do so by teaching their children what we teach ours?

At one time we used to answer this question in an unwavering affirmative. During the hundred years when Great Britain ruled in India, we used to educate such Indian children as we *did* educate according to a curriculum identical to our own. Not only were they instructed in the 93 elements of the so-called 'Periodic Table', which lists all the different stuffs on earth in order of their atomic weights, but they were also made to read Dickens' descriptions of a mid-Victorian London they had never known and, later on, to sing the incomprehensible satires of Gilbert in the unfamiliar rhythm of Sullivan. But did this mean that science *was* just science, and that, if people from far-away places were prepared to learn it and take it home and use it, all they had to do was to stay and get a PhD and all the rest would follow? In no time at all, the buses would start to run in the bush and the local citizens could start to work in the plastics factory as soon as they chose.

Surely the joke was just a joke when it told of the active colonist in his soki topi trying to persuade the indolent man lying dozing in the shade to get up and work so that he would become rich and prosperous and could therefore after a lifetime of effort make himself wealthy enough to be able to afford to lie indolently dozing in the shade?

Or could it be that our own culture has begun to change since then? Perhaps even the beliefs of the early scientists and engineers were derived from the time before Einstein was born when physics, chemistry and mechanics, complicated though they were and hard to learn, were – or so it seemed – basically simple. But now that Einstein *has* been born, even science has changed. Relativity, the new kind of thinking that he invented, means that we have to look at basic things, like the weight of a sack of potatoes – or an atom – from a point of view which depends on where we are and how fast and in what direction we are moving in relation to the potatoes (or the atom). And perhaps we can now see that the same principle of relativity also holds for science itself. What it is and what it means to any particular society is affected by the relative velocity with which that society is moving and changing as it moves with the fluctuating stream of history. And this stream moves at an uneven pace not only for the same community at different times but for different communities at any one time. There were centuries when things remained virtually unchanged in China and remained static in Japan. There have been long periods when progress has advanced steadily, bit by bit. And there are other periods when abrupt acceleration takes place and suddenly, after ages spent using bronze, a community takes to the use of iron. After centuries when the speed of a horse has seemed to be fast enough, the cultural tension adapts itself to nothing short of the velocity of a steam locomotive or a bicycle.

The relativity that Einstein thought about referred to solid bodies, whether they were atoms, tennis balls or stars, and to time and motion in space. But now it seems – even if in a somewhat different way – to hold also for events, such as the understanding and utilization of science. What science is and what it means to any particular society is affected by the relative velocity of that society as it moves through time (assuming that it does move at all) from point A, which for convenience we once used to call barbarism, to point B, the pleasant environment in which people of culture dwell. The man with a PhD has a different weight in a culture of advanced technology than he has, let us say, in India or Botswana. The science he knows at Harvard or Reading is the same science he takes with him when he goes to Kinshasa or Khartoum, but, through the workings of cultural relativity – if in all humility as a minor Einstein I am allowed to invent such a term – it does not act the same when he gets there.

It may perhaps clarify the idea of the cultural relativity of different societies each travelling along time's pathway at different speeds and in different directions if we reflect for a moment on our own society at different stages in its cultural journey. In 1785, we find a great man, the Hon. Henry Cavendish, eccentric, enormously wealthy but entirely unworldly, discovering the composition of water to be H_2O. This, though of enormous interest to those few scholars who understood what Cavendish was talking about, left Western culture totally unmoved.

A hundred years further along the cultural orbit in 1879, we come upon Thomas Edison. Edison was not so much a scientific originator: he did not uncover previously unknown facts about the universe. He was, however, an originator of a different kind and as such was in advance of his contemporary culture. He introduced the notion that practical economic things could be done by a man like himself; while not necessarily gifted with a talent for scientific discovery, he could yet understand the scientific knowledge of the day and identify those areas where some new piece of knowledge could be practically useful. Edison was one of the first people to hire other scientists to work on some specific problem in what we should describe today as 'R and D' – research and development. His great triumph was not to illuminate men's minds, but to perfect a reasonably long-lasting electric lamp bulb and thereby illuminate their houses. In doing so he made a profit for the businessmen who had financed his operations at Menlo Park. Edison came to the verge of bankruptcy on the way, but he succeeded – and Western culture was never thereafter the same. The science of Edison (or was it technology?) achieved one thing in 1879; the very same science would have been much less potent among the broughams and coaches of Cavendish's eighteenth-century London; it would have been almost completely impotent among the men of culture of Newton's time, when Sam Pepys was driving back to supper across the ruins of the burnt-out City, with his Home-Guard sword in his hand ready to repel muggers.

'Ride a Cock-Gee to Banbury T,' sang Aldous Huxley, 'To see a fine bathroom and W.C.'. His parody of a nursery rhyme was forecasting what our children would recite in the Brave New World that science was bringing. Yet the parody cuts deep. The man of twentieth-century culture has sacrificed a good deal for his plumbing and electricity. The indigenous inhabitants of Paris, London and Boston must, no matter how reluctantly, enjoy their life of high culture in the knowledge that a significant number of their fellow citizens are homeless because the cost of housing – that is of houses with knobs and bulbs and taps and switches and all that go with them – is high. The scientific culture, being based on an implicit assumption that the application of science can achieve anything, has up till now assumed that it can make people richer too. Until very recently it was assumed that there was no upper limit to how rich a community belonging to the science culture could become. And even now, when the Club of Rome and the authors of *Limits to Growth* have cast a doubt on this, people do not in their hearts really believe them. In the official form which travellers aspiring to visit Saudi Arabia complete when they apply for a visa there is a space in which each applicant is required to specify his religion. Should a liberalized Westerner write 'none', he does not get his visa. As the Saudi Arabian officials put it: 'All sane men must submit themselves to God in one or other of His forms. A man who denies His existence must be mad. We do not want madmen in our country.' Deeply ingrained in our modern technological culture is a belief in the infinite capacity of science. It takes more than a couple of pessimistic dissenters to change this conviction.

In the community where the scientific culture is most deeply ingrained – I refer to the United States – it is widely accepted that unhappiness is not, as other cultures may have believed, part of the human condition. Faith in science leads to the belief that it is a condition like smallpox or pneumonia that is amenable to scientific manipulation. Whether it is due to a man's quarrelling with his wife; to feeling inadequate in his business; or to fear of the impending political situation; he will be convinced that cure for his melancholy and restoration of a state of contentment and joviality will be ensured by the science of psychology as applied to the individual state through the medium of a psychologist. Split open a skull in Rhiad, the capital of Saudi Arabia, or in Albany, the capital of New York, and you will find the same organ inside: the physiological and biochemical mechanism of brain whether in New York State or in Saudi Arabia is the same. Yet the science of psychology – at least that part of it inherent in the practice of psychiatry – will produce different results under the different cultural circumstances of Arabia and Scandinavia.

The strange belief in the unbelievable which has emerged as part of the 'scientific' culture is in several respects a paradox. The gospel of scientific method from which faith in science grew purports to be precise, rational, quantitative and based on exact measurement and observation. Yet the results which could be expected if the reasoning were rational and the facts factual do not occur. The second part of the paradox is that, while the belief in science – like the adherence to other and earlier official religions – is part of the current orthodox cultural pattern, underneath and in their hearts many people still adhere to much older cultural mores. To go to a psychiatrist is in some respects not much different from taking a Turkish bath. A bath may make a bather feel better, yet he does not seriously need to believe that it has much effect on his health. Similarly, to visit a psychiatrist may encourage a transient improvement in cheerfulness but whatever it is customary to claim in public, one need not necessarily believe that it has caused one, in fact, to change one's mind.

In an earlier culture, monarchs and their ministers would declare certain days as days of prayer when the nation would gather together to appeal to the deity for rain, or possibly for peace. Devout people would put their hearts into such prayers and the ordinary people would join, with greater or less fervour, in support of the cultural exercise. Few people, looking back on such events, would care to assess the hard evidence showing what effect these occasions had on either the meteorological or political climate of the times. Today, in the cultural atmosphere of science, there have been occasions when calls have been made, not to God, but to sociologists for guidance in solving the political troubles of an oil company dealing with an intractable government or in handling student unrest in universities, colleges and even children's schools. Again, the practical results of these scientific exercises are as difficult to assess as were those of the earlier religious exercises.

Has there been cultural progress in the way technologically advanced nations have moved on from a general acceptance of the tenets of Christendom? Christian beliefs and canons of behaviour found no contradictions in parallel acceptance at various times of the social necessity of capital punishment; of decapitation, garrotting and the extreme operation of the rack; even of burning at the stake. Today general approval of foxhunting is assumed by half the population, particularly of urban thinkers, to be the extreme example of turpitude not only because of the extermination of foxes but more particularly because foxhunters enjoy the chase. The man of culture in a scientific age can more readily accept as good behaviour the destruction of such animals by such scientific means as asphyxiating gas of appropriate molecular configuration. But the progress of culture from a classical and Christian age to one of the rationality inherent in science shows signs of deviation from a straight-line progression towards more and more 'rationality'. And does belief in the ultimate omnipotence of science extend not only to the control of disease and the manufacture of machines, but also towards the rational mating of men and women? Should there be rational selection of genes for the construction of perfect children to be educated under ideal conditions of rationality to form exemplary members of the 'free world' or the 'socialist nations'? There are signs that the man of culture will not accept that we go quite so far.

One of the most significant signs that the fundamental basis of the scientific culture was changing came from within science itself. On 1 February 1976 Werner Heisenberg died. As Edward Teller, one of his pupils, pointed out in a perceptive obituary notice, Heisenberg, together with the great Danish scientist, Niels Bohr, was mainly responsible for bringing about a virtual revolution in the philosophical basis of science. Heisenberg would anyway have obtained as much celebrity as any ordinary scientist could deserve for his development of quantum mechanics as a whole new branch of physics. Important as this was, it was an advance along the lines of conventional scientific philosophy; once discovered, it might have been taken as one further example of the power of rational thought to overcome the secret mysteries of nature. Certainly it further increased the ability of men to exploit the power and wealth gained through engineering skills and scientific knowledge.

Heisenberg in fact broke away from this line of thought.

The old reasoning, puzzling over the nature of light, had led to the deduction that light could *either* be produced by a series of particles, whizzing swiftly from one place to another and bouncing like ricocheting birdshot from solid reflecting surfaces, *or* could be explained as rippling waves flickering throughout the body of the void. It could be *either* a particle, *or* a wave; but if it were one it could not be the other, could it? Heisenberg, seeing further than his fellow men, argued that it could and proved how it might be so. This hypothesis, new to science but old to philosophy, he called 'Complementarity'. For the first time scientists, men and women brought up in the orthodox culture of their age, had to stretch their minds in a new way. They became aware that complementarity, the ability to believe several contradictory things at the same time – a process frequently necessary in real life – had now become an essential feature of the scientific culture, which up till then had prided itself on the potency of its strict rationality. Lewis Carroll had foreseen what Heisenberg was now proposing. When Alice objected that 'One can't believe impossible things,' the Queen merely replied, 'I daresay you haven't had much practice ... When I was your age, I always did it for half-an-hour a day. Why, sometimes I've believed as many as six impossible things before breakfast.'

Heisenberg demonstrated to the men of the scientific culture that two contradictory hypotheses could both be true at the same time. He showed that there might be a limit to the accuracy with which an observation could be made beyond which it was, not only technically, but philosophically, impossible to pass. For example, the more accurately one fixes the position of a particle at any particular instant, the less accurately can one measure its speed at the same instant. This proposition is now called Heisenberg's 'Uncertainty Principle'.

It is surprising – as the discoveries of great men so often are after they have been made – that the orthodox scientists of the pre-Heisenberg age of reason had overlooked this principle. The facts upon which it is based were, after all, well known in general terms. An obvious example, ignored before Heisenberg came, is that of the gnat, winging its way northward towards Scotland at a steady 10 mph which runs into the locomotive drawing the crack train, the Flying Scotsman, travelling southward at 80 mph. The position and speed of the gnat 1 cm before it strikes the front of the locomotive can be measured with some accuracy, as it can when, the train having moved 8 mm further south and the gnat 1 mm further north, it is 1 mm away from the locomotive. But while the position of the gnat can be fixed with considerable precision at the precise microsecond it strikes the locomotive, to measure its velocity at that instant is a very much more difficult problem.

The new kind of thinking introduced into science by Heisenberg, developed to cope with problems of physics, that most exact of the exact sciences, brought with it cultural overtones wider than those of science itself. Complementarity is not new, it had merely been temporarily out of fashion. In an earlier culture people used to ask the question: What am I? Am I a biochemical mechanism, or am I an immortal soul temporarily encased in an envelope of flesh? In fact Heisenberg applied his scientific principle of complementarity to a wider sphere of observations than those of science. He extended it to some of the most important matters with which a man of culture is concerned, among them the problem of people's behaviour towards one another.

While Heisenberg developed his new thinking into the higher reaches of cultural activity, his uncertainty principle was spreading out to affect some of the lesser cultural assumptions of his times. Let us, for example, examine one of the more popular beliefs of a scientifically cultured society, that science enables us to attain absolute safety.

The prophet of this particular gospel was Ralph Nader. Nader, it will be recalled, first rose to prominence by his attack on General Motors. He accused them of making motorcars which he described, in a publication with the same title, as being 'unsafe at any speed'. This campaign as it gathered momentum became a general attack: every product of science-based technology, not only automobiles, was seen as carrying with it some degree of risk to the user. While this was philosophically unassailable, the reaction, which in an earlier culture would have been resignation to the fragility of human existence and thanksgiving for each day's safe survival, was quite different. It stimulated a great religious revival in men's faith that science could save them. In the current culture one former belief became a heresy so profound that it was barely contemplated even by the most abandoned of wretches: that every human activity carried with it some measure of danger and that one should – in the words of the ancients – 'call no man happy until the day of his death, he is at best but fortunate'. On the contrary, it was assumed as just by the cultured men of science that there be laid upon every manufacturer a solemn duty, to apply the power of science and technology to the fruit of his labours to ensure their absolute innocuousness. Thus everybody, whether he was a manufacturer of motorcars, cosmetics, food, or a supplier of medical services, was to be compelled to adhere to standards of behaviour laid down by the community. These standards, based on scientific knowledge, were to be a talisman, a protection against all ills.

It is clear that societies which have progressed less far along the road towards the culture of science believe in the attainability of safety from the world's ills less devoutly, but even they will take reasonable pains to guard against preventable ills. It all depends where the line of preventability is drawn. The Bridge of San Luis Rey, suspended across the great gorge in South America, lasted for years until it fell carrying to their doom the diverse people whose lives had brought them together to cross at the fatal moment. It is as sensible for a scientific community to vaccinate its people against smallpox and guide aircraft by radar to a safe landing as it was for their ancestors to build safe bridges and erect lighthouses on the rocky promontories around their coasts. But the criterion of safety is moved further and further along the range of life's dangers, seeking preservation against more and more minute possibilities of harm: destruction of the earth's atmosphere by hairspray; fracture of the elbow by resting it on the open window of a motorcar (an Australian regulation exists to safeguard the citizens from this); until the doctrine of absolute safety itself comes to be challenged.

The challenge arises in two areas. One reaches a point when to extinguish the last vestige of danger, if such were possible, would require a degree of disruption of normal human existence that caused more damage than the harm it was designed to prevent. This is the point in Heisenberg's uncertainty principle when the position of the very, very small particle is being scrutinized with such intentness that its speed and direction of travel are lost. I foresee that the effort to eliminate the ultimate risk of cancer and heart disease may be such a point, that the final endeavour to ensure that old people when they die at least die perfectly healthy, may suggest to the men and women of the scientific culture that they can no longer determine the direction their society is taking.

Crossing the Atlantic by sea in the most advanced passenger liner propelled by up-to-date steam turbines, the modern traveller looks up at the exhaust pipe – which in the earlier days of steam ships was once the funnel – and, seeing a wisp of vapour against the sky, cries, 'Ho! Pollution!' And the ship is alone on the ocean where once the navigators crossed and recrossed, the lookout on each foreseeing the advent of his fellow mariners by the smudge of smoke (and it was smoke) on the horizon. When Heisenberg talked of the uncertainty principle, he referred to a single photon of light and a single particle of matter, both so small that they could only be seen by the trace they left within the instruments constructed to give evidence of their existence. But uncertainty is well chosen to describe the spirit of the citizens of the science culture that has come to the advanced nations of the twentieth century. There is an ever-sharpening focus of the search for harmful ingredients, in food, in paint, in insecticides, in the materials of construction; examine glass with sufficiently sensitive analytical probes and a toxic substance will be detected (if not lead, silicon); few if any plastics could survive the degree of scrutiny that modern tools can now provide; the rain drumming on the asbestos roofing tiles will knock free at least one fibre that could, if conditions arose (tomorrow or a century hence) do harm, no matter how little, to someone; any machine that can be designed may fly apart and do damage . . . in the end we can show everything on earth to be harmful. In our cultural uncertainty, the fable of the Sleeping Beauty has come true. No matter how secure the castle in which the King and Queen confine their darling daughter, in the end the Wicked Fairy's curse will come about: the Princess will prick her finger and fall asleep for a hundred years.

Parallel with the search for perfect safety is a belief that has gradually grown up: should harm occur, someone must always be to blame and thus liable to be sued for restitution. As scientific understanding advances and it becomes possible to detect smaller and smaller causes of more and more remote hazards, it is just possible for those who produce goods or provide services to keep ahead of events if they try hard enough. Not long since, a sufferer and his family knew the fallibility of the doctor's medicine. Patient and physician were aware that all men suffer and die and that the power of the healing art was limited. Today, the art of healing has become the science of medicine. There is consequently an assumption that a competent doctor, who can control the infection of wounds and the very existence of smallpox as a disease at all, can be held responsible for *any* suffering or harm. We are virtually in sight of the state of affairs when a patient's death can be taken as a sign, not of God's will, but of the doctor's incompetence for which he can be sued. Shortly, no doctor will treat any condition no matter how trivial without first commissioning a battery of biochemical tests, of x-rays, of psychiatric data: all not to protect the health of the sick man but to protect the doctor against litigation. A boy might once, before the age of science, fasten four perambulator wheels to a wooden orange box and make himself a toy truck. Not any more. If he should lend his homemade vehicle to a friend who broke his leg in it, the case would be actionable. The truck must undergo scientific tests and be modified a hundred ways in an attempt to indemnify everyone concerned with its manufacture.

Even today, a food manufacturer if he is prudent will take pains to ensure that his products are *not* analysed. Time enough to undertake the exhaustive and expensive tasks necessary to demonstrate the harmlessness of the chemical components of sponge cake, kippered herrings or fried chicken *after* they have been identified. If nobody knows what the ingredients are, nobody can demand that their safety should be established.

There is one result of this commendable pursuit of safety, corrupted as it has been by the mistaken belief that science can certainly allow the cultured citizen to attain a safe, sanitary and assured paradise; or, at the worst, if the guaranteed satisfaction is not provided there will always be someone to sue. Life, far from becoming satisfactory, becomes impossibly expensive. The safe motorcar is expensive to buy and correspondingly expensive – due to its pollution-free combustion system, its pneumatic cow-catcher and its interior-sprung safety harness – to run. Children's toys are only within the reach of the rich or the profligate owing to the cost of producing objects that cannot be sucked, swallowed, used as weapons or projectiles; cannot pinch, scratch or run a child over – cannot be dropped on baby brother's head or pushed into his ear. And if medical help is required, no one can afford it, neither the State nor the patient. Even in the culturally imperfect days of the 1970s, doctors in the United States, where the most devout believers in the omnipotence of science dwell, could no longer afford to pay the enormous insurance premiums needed to cover the gigantic damages awarded against them should one of their patients' diseases take an unforeseen course. And this was before the death of a patient was adjudged to be a proper cause for legal action against the physician who treated him. Houses, clothing, medicines and drugs, domestic appliances – all became more elaborate and expensive as regulations for their safety proliferated.

The culture of our age at first seemed so desirable. As it matured and then, as some people judged, began to decay and become effete, here and there were those who sought an alternative society. Many of these alternatives soon proved to be unsatisfactory. For middle-class citizens accustomed to their cars, television sets and central heating whose lives had been supported by doctors, lawyers and the man who came to fix the telephone, life on a muddy fifteen-acre farm in the Outer Hebrides soon proved to be for them as laborious, unrewarding and hard as it had for the original inhabitants – who had been only too glad to depopulate the Scottish Islands as soon as a job in a plastics factory in a new town to house the Glasgow 'overspill' became available. But there is a need for some alternative to our present direction of travel in the technological society. If I were asked to invent a breakaway culture to reform, and perhaps renew, the one that has grown up and is still growing, I should construct it like this.

A community prepared to accept the fact that medicine, even when supported by science, possesses only limited powers and that everyone must die, would agree to accept that doctors are prepared to do their best as honourable men: we would agree not to sue them. And if the baker was ready to eat his own bread and give it to his own children, his customers would eat it too. The chance of anyone being infected by unwrapped bread is infinitesimal; wrapping is merely an example of modern prudery not unlike that of our Victorian ancestors who fixed little skirts to the legs of their pianos and dining-room tables, so offensive did they deem the sight of bare limbs. What baker puts his bread on the racks without touching it? No, he would not wrap it. Nor would the purchasers take him to court if it were stale. They would, of course, reserve the right to grumble: if they found a cigarette butt in a loaf, they would grumble very much and, unless the baker had a good excuse, take their custom elsewhere. In my new society there would be sub-standard houses with the wrong number of cubic feet of space and do-it-yourself plumbing; in such housing those who opted for the community would live with open eyes.

It may perhaps be argued that I am advocating mixed monasteries in which the whole family can live side by side with the world of the science culture, eating simple meals, wearing qualified (and non-guaranteed) hair shirts carrying no certificate of non-shrinkability.

It has been justly said that the United States Food and Drugs Administration has caused more needless alarm and apprehension among the people of the science culture, not only in America but throughout the entire technological world whose bible is the *Codex Alimentarius* than have a hundred sensational journalists and broadcasters. So desperate are they to show the supremacy of Man the Scientist, in whose loving arms all men can be guaranteed safety, that they have cast suspicion and doubt on colours that make food gay; on flavours that make it delightful; on new foods; and increasingly, as they stuff their experimental rats with familiar comestibles until they squeal in anguish and one organ after another breaks down with the strain, on long-familiar articles as well. Since water intoxication is a well-known syndrome, it is a miracle that they have not taught more people than those few who eschew it in favour of strong drink to develop hydrophobia – the fear of water.

When our man of culture changes his ideas, he will drink water, eat his 'peck of dirt' – as my grandmother used to say – and use the fruits of science, but use them judiciously as a humble believer in something other than his own cleverness, useful and comforting though this may be.

9 There and back

After finishing the last chapter on a high and philosophic note, I ought to set down in more precise terms how ordinary people might so arrange their affairs as to live a comfortable and civilized life with the science and technology that gives us such remarkable advantages over our predecessors.

As I write, hard-headed members of the radio and television industry are organizing themselves in a matter-of-fact way for a future which they expect to see in ten years' time or less: when the television set in the corner of the living room will not only be used for a few hours of entertainment in the evening. Data banks are already being set up to allow every citizen, by paying only a few pounds more than he already does for his television licence, to press a button on his 'View-data' adaptor and read his letters off the screen without the need any more for paper and envelopes, stamps or pillar boxes. He will also be able to read the newspaper without having to go out and buy one. Thus the judicious use of science and technology will do away with stationery and post offices, make postmen unnecessary, avoid the need to cut down all those trees and turn them into newsprint, and allow all the printers who so obviously find printing newspapers in the middle of the night disagreeable to go to bed like the rest of us.

The same machine – not in a remote space-age future but almost within the lifetime of the present government – will save all the gallons of petrol that salesmen now use in driving round the country to call on their customers. They will be able to show their wares on the television screen to potential buyers and answer questions about them as well. There is still some debate as to whether the process already being tested by the BBC will make it unnecessary to build any more schools, either comprehensive or otherwise; but the time at present being spent by the teachers in teaching and the pupils in learning can undoubtedly be done better by the personalized and computerized lessons transmitted to the home, allowing such schools to continue to function on that aspect of what they conceive their work to be in relation to the social and recreational needs of the pupils. And already the boom area in the electronics industry is the burgeoning sales of television games.

But I need hardly continue by enumerating the varied activities, such as the banking system, the stock exchange, much of the time-consuming tedium of the doctor's surgery, and the expensive worry involved in taking counsel's opinion, which this single development in technology could already bring to an end by linking the domestic TV screen to the appropriate computer data banks without any need to wait for future advances and discoveries. The fact is well established that science and technology as utilized by a community enrich life (if we use the word 'enrich' in its common meaning) by providing the various attractive goodies that I have been enumerating in the course of my narrative. Science and technology also lead to that strange, loaded, pejorative conception which haunts industrial citizens, namely 'unemployment'. Virtually every technologically advanced community finds an increasing number of its members, as the saying is, 'unemployed'. This does not mean that they are doing nothing. It merely implies that they are not for the moment involved in the various manufacturing operations or in the procedures by which they are serviced by which some of the economic indices of the community are measured. It is a matter of current belief that such 'unemployment' is bad, that people thus 'unemployed' are to be deprecated because they are not producing 'wealth', and that in due course the 'unemployed' people will become 'employed' when some so-far unspecified action is taken. Reason, however, makes it hard to avoid the conclusion that unemployment will never diminish. Have I not just described how one existing development in a single industry can render postmen, printers, salesmen, filling-station attendants, workers in commuter-transport systems, teachers, drivers of school buses, and tree-fellers – to name but a few – unnecessary? It is true that unemployment, in its meaning of nine-tenths of the community working for 35 hours a week and one tenth not working at all, could be made to disappear (while, in fact, remaining the same) if the whole community worked for 31½ hours a week. But even though this is arithmetically so, what is equally undeniable is that the ever-increasing ability of the technology which we invent and install for the very purpose of doing industrial things, does do them and, in consequence, leaves less for people to do.

There is no real need to worry, however. All that is needed is to devote some of the rational thinking we apply so lavishly to science on the way we organize our happiness. It is true that rational thinking is difficult, but there is really nothing to stop us doing it, nor does it require the consumption of scarce raw materials. In the matter of food and nutrition we are already beginning to glimpse the road to happiness. Gradually, it is being understood that all the wonderful science that makes not two, but two hundred and twenty-two grains of corn grow where only one grew before *does* enable us to produce all the food we need to feed the starving millions. All that is required is rational thinking to find a way to share the food out. Science and technology have solved the problem of producing the world's wealth as well. In addition – and this is why it is justifiable to be so sure that we're winning – there are signs that we are beginning to understand what wealth is.

Perhaps the most illuminating comments on this topic were made by Hugh Stretton of the Australian Broadcasting Commission in 1974. People tend to be unhappy when they are dubbed unemployed because they feel that they are not producing wealth while they stay at home, even though everybody knows that it is the inexorable advance of science and technology that sent them there. This sadness at not having to go to work was not always so. The earl in his castle and the coal-owner on his country estate never worried. Indeed, they used to get satisfaction from the happy thought that their extravagant habits gave employment to the working classes. How this age of science has caused things to change! But Hugh Stretton shows us one of the avenues by which we can regain our happiness. Social workers, home helps, local authority administrators and canteen supervisors are all content to know that the money they earn – often after long and difficult periods of technical training – contributes to the gross national product. The unemployed can equally share this virtuous happiness and overcome the melancholy of the old-fashioned pre-scientific work ethic – particularly if such unemployed are housewives, as most of them are.

Visualize a hundred married ladies, each cooking dinner in her own kitchen: a burden on the national economy. All this can be changed by the power of thought. The lady in house No. 1 goes into the kitchen of house No. 2 and cooks the dinner there for No. 2's family. For this she charges No. 2 full restaurant rates. No. 2 has no difficulty in meeting the expense because *she* has received the same amount for cooking No. 3's dinner. By the time No. 100 has finished washing the dishes in No. 1's kitchen and taken off her apron, the gross national product has received a distinct fillip. More important than this, however, is the fact that a hundred housewives are happy at having absolved themselves of the stigma of unemployment without actually doing anything different from what they did when they considered themselves unemployed.

And the ingenious Mr Stretton goes further. Why should not the people at No. 2 let their bedrooms to the people at No. 1 at full hotel rates, fix No. 1's bath-tap washers at plumbers' rates of pay and mend their fuses at the rates charged by electricians? The money so earned will serve to pay the family at No. 3 for renting their bedrooms. All down the road the children – ever quick to learn in this sophisticated technological age – would be renting each other's backyards to play in and hiring each other's big sisters at the official local-authority rates for play supervisors. I forbear to follow Mr Stretton down the intricate paths of academic economies where he foresees bankers, studying the phenomenal growth in the national economy arising from these amiable and harmless manoeuvres, rushing to invest in activities showing so fabulous (and I use the adjective advisedly) a growth in capital-output ratio, sinking their money in double beds.

This parable has not yet become reality, although there seems to be no reason why it should not. It is, however, a glimpse of how people can win the contest with the system they have devised. One of the early signs that we are beginning to apprehend how a modern industrial society, with its advanced industry making use of the matchless powers of science, can be managed is the fact that we know that Mr Stretton's amazing ideas must be taken seriously. It is a sign that we are winning, that we are beginning to regain our self-confidence and to recollect that it is we who are using science for our content and that we need not change our lives to serve the scientific ideas and the scientific machines by which our industry runs. On the contrary, the ideas and machines are only worth having if they serve our humanness. Although science is knowledge of the immutable truths of the universe and takes no more account of human wishes than does the wind, the machines and processes that make use of science are only constructed for that purpose. We use the wind to drive the ship in the direction *we* want to go. And if the computers that operate the banking system are programmed to classify as 'wealth' the work a strange woman does in looking after an incontinent old lady and to classify as 'consumption' or 'unemployment' the very same task when it is done, not by the strange social worker, but by the old lady's daughter, there is no reason why we should not change the programme if we want to.

Quite a number of people have begun to point out some of the oddities of the way in which the gradual development of scientific ideas has caused what might be described as professionalization gradually to spread across a wider and wider area of human activities. The Wright brothers made the first aeroplane in a garage. Only professionals in the 'aerospace industry' make aeroplanes now. And only qualified chemists and engineers could hope to make nylon socks; spinsters spinning yarn could not compete any more. Nor is a mother expected to teach her child to read: a professional is paid to do that today – or should I say 'was paid yesterday'? As the tide of technology advances, more and more of what were once the ordinary activities of life become professionalized. To read in Jane Austen and Dickens of devoted sisters or husbands, watching with loving anxiety by the sickbed of a relative, has become the ethnological study of a distant primitive society. Rightly, we glory in the antibiotics which science, in the hands of brilliant men, has provided to prevent our loved ones dying as young as did those of our ancestors; but there are few of the sick, as well as of the well, who will not fumblingly have felt something missing from their lives at having no part (except through their government and their taxes) in the care of a beloved.

The tide of technology has advanced across the lives of those whose skill has made it possible. It has advanced so far that it seems to have left nothing untouched. The genetic code of DNA, the chemical molecule of heredity, has been mastered and conception therefore is fully under scientific control. Death is no longer a pale reaper, tapping a man on the shoulder when his hour has struck. It has become a legal argument to decide whether and under what circumstances the pacemaker and the resuscitator, the pump and the oxygenator are to be switched off and by whom.

Even after death has been certified and the chart of the electroencephalogram safely filed to ensure that no subsequent legal suit will be brought, the last froth of the advancing tide reaches still further forward before it begins to ebb.

A balanced psychiatric approach to the mental health of an up-to-date community would deprecate too close an attachment between individual citizens. Dying of a broken heart, though it rarely occurs in modern industrial communities, is to be discouraged. Should instances be found to occur at all widely, appropriate administrative measures would need to be taken by the public health authorities to relieve the tension. It is only recently that new entries to the public health service have been properly trained to deal with death. Not only have appropriate criteria been established to identify the exact moment of death, which is, as it was in the past, a matter of administrative importance; but also to determine the point at which the death of the individual occurs without compromising the viability of organs useful to the community as material for transplantation – the kidneys, the heart, the eyes, skin and bone marrow, and perhaps the brain. This is obviously a matter of some philosophical subtlety. But medical science has shown signs of extending even beyond death. Consideration is now being given to the eradication for the public good of any vestiges of mourning that yet remain among the more backward sections of the community. Grief is under examination, not as the response of a feeling heart to the breaking of a deep attachment, but as a pathological symptom to be overcome, should all else fail, by sedation.

This is perhaps the point at which the tide is beginning to ebb. Here and there change is already apparent. Is there, I wonder, an indication to be found in the unsafe society of the old?

Old people are in a particularly good position to assess how best a modern technologically based society is running. After all, it is they who are the beneficiaries; they have come into their inheritance which it is the purpose of the whole system to provide. The young must learn so that they may grow up to take their part; full-grown they work to increase productivity and stimulate economic growth – which somehow or other never seems to attain its full peak; later as managers they worry about maintaining the national prosperity, which when they finally retire – that is, when they are adjudged to be old enough – they can look forward to enjoying.

Up till now it has been fashionable to be apprehensive at the increasing proportion of pensioners in each advanced industrial community. The increase is, of course, partly due to the scientific successes which enable more and more people to attain their full life span without dying of infectious diseases beforehand. Soon the numbers will increase still further as one by one the degenerative diseases, of which the various cancers are the most noteworthy, are overcome as well. But a more potent factor underlying the growing numbers of pensioners is the decision of the community to include more people within this fortunate group by allowing them to join at a younger age. A. P. Herbert once wrote a satirical parody of the lyric of the Mountains of Mourne. He asked:

Oh, won't it be wonderful after the war
When there won't be no rich and there won't be no poor
We won't have to work if we find it a bore
And we'll all have a pension when we're 24.

He then went on to point out, quite rightly, that

There won't be no sick and there won't be no sore,
The beer will be thicker and quicker and more.
But there's only one avenue I want to explore,
Why didn't we have this 'ere war before?

As so often happens, the satire of one age – and Herbert wrote in the 1940s – can become the truth of the next. Great companies already retire their officials at 55. The effectiveness of science-based technology is such that the community can virtually choose any age it pleases. In doing so they are not, as glib commentators so often wrongly point out, laying increasing burdens on the shoulders of the young, for pensioners do not do nothing when they retire. We are all of us waiting to die, and the old have no more wish to do it than their juniors. The supreme advantage that is enjoyed by pensioners as a group is being free to look around them, to reflect, and to carry out whatever activities they choose under a different set of rules from those which the rest of society obey.

There are ingenious bodies of pensioners in London and elsewhere who have organized themselves to do all sorts of things which ordinary people, fixed to the main technological system, cannot easily get done. Making use of a unit of currency based on the value of one hour's work, the group can get clocks mended and home-made cakes baked; can afford plumbers, electricians, plasterers and paper-hangers. As the group grows, they can achieve almost anything, from teaching French to growing cabbages. Here then is a microcosm for the future. On a foundation of a 'pension' provided by the wealth-producing capacity of large-scale technology which, once it is established, requires little involvement of people's time (that is, of their lives), the pensioners' group are inventing a social system of their own. It is in no sense a primitive operation. At the centre are people acquiring their 1-hour units of currency by matching the day's accumulation of needs, which have been recorded on a battery of telephone-answering machines, with the services that are on offer. The potentialities are limitless. The basis of mutual assistance, which becomes eroded in the larger official world, allows for transport at economical rates since, between members of 'the club', the expense and complexities of transport regulations do not seem to apply (particularly if nobody mentions the matter). Nor does the member baking cakes provide a list of ingredients with their vitamin contents, maintain a laboratory staff to certify the moisture content or a date after which the product cannot legally be eaten.

The strength of the pensioners' position, which gradually they are beginning to appreciate, is that if, having evolved a system which is in some degree an unsafe society, they actually suffer from eating cake after its 'open date stamping' has elapsed, they know that they will not be eligible to sue for damages – or, if they are the cake-makers, be sued. In

America, there are similar 'unsafe societies'; medical service, like plumbing, being out of economic reach, a member consulting a doctor whom he recognizes as an honourable man and a member of the group is aware that he will be unable to bring a suit for malpractice if he does not get better – or even if he dies. Nor can another sue if, being transported in a fellow member's car, a wheel comes off; or when painting someone's house, the ladder slips.

I sketch the idea with a broad brush, the details I leave to evolve as time passes. The advantages are two-fold. 'There is nothing either good or bad, but thinking makes it so', said Shakespeare. In my unsafe society all members would know that they were banded together for mutual advantage and support in this hard world. If the worst came to the worst and something went wrong, knowing that you had all done the best you could without the need for one legislative restriction piled on another to compel you – well, as a pensioner you'd have had a good run!

As the group grows, as the pensionable age creeps lower, as non-pensioners, appreciating its benefits, clamour to join as honorary members, it will suddenly become apparent that the whole lot of us *are* banded together for mutual assistance in a chilly world, and that we *are* all pensioners one way or another with only a comparatively short time to live. Looking at our affairs from this point of view, I see little difficulty in believing that there is no reason to worry about the direction in which we (there is no one else) decide technology is to march.

Mark you, we do not exactly decide where we go even though the signs are that we are moving towards a lovely goal. Official plans are designed to ensure the production of the right number of motorcars, the correct tonnage of plastic sheeting, and the appropriate hardness of the currency. Yet, as I have just described, the plans do not exactly fulfil the norms they are designed to achieve. Suddenly people appreciate that the cosy little houses they once denigrated as slums had certain attractions as human habitations and that even well-designed under-floor-heated council houses can be bent to conform to the irrational pleasures of their inhabitants when the people who live in them have erected a distinctly unsafe shed in the back garden.

In the same way that untidy human instincts produce a somewhat chaotic unsafe society, so also does technology itself, ostensibly so serious and rational, unexpectedly bubble and sparkle in the unquenchable warmth of the human spirit. Consider the serious topic of food technology. Serious it is, yet of the thousands of food scientists who labour in the laboratories of the food industry in all its diversity – in the margarine factories and the plant bakeries; in the great mechanical units producing sugar confectionery; in the soft-drink bottling halls and the breweries; in factories where they make meat pies; and in the newly constructed units where, for the first time in human history, nitrogen, deeply cold and liquefied from the impalpable air, is poured over hot roast beef and apple pie alike so that, combined as meals, subjected thus to the 'cook-freeze' process, they may be filed away, to be revived in forced-draught convection ovens and served piping hot six months later as good as new – in all these places, out of every thousand diligent technologists, nine hundred and ninety spend their life in pursuit of aesthetic goals. Suddenly, they will awake from their sleep and, more numerous than poets, painters and violin players, they will take their place with them in man's worthy search for beauty.

Today, there is a hue and cry in the hunt for 'additives' and concern for their innocuousness. But these 'additives', what are they? 'Improvers' aim to produce the even and inviting texture of Mechlin lace in every slice of bread. Great treasure is spent by the relatively impoverished society of Great Britain to import wheat from Canada containing protein of a particular 'strength', not because it is more nourishing than other wheat, but because when it reacts with the chemistry of the improver it forms a crumb structure as uniform as gossamer. No gap is ever to occur in the fragile porcelain structure of the bread to trouble the eye of the connoisseur in its transient life between the toast rack and the teeth, nor ever shall marmalade fall through an adventitious hole to make sticky the trousers of the consumer at his breakfast.

The subtle chemical configuration of emulsifiers in bottles of orange juice is designed to maintain the translucency of amber in the fluid with never a trace of a disfiguring oily ring between wind and water. Organic chemists, poring over their microscopes in university laboratories are studying the crystal structures of the palmitates, the oleates, the linoleates and the linolenates – magic words to conjure with magic crystal structures from oils once used in lamps all over the world. And why do they do this? They do these things, combining skill, knowledge and intuition, in order that the firm smooth structure of the golden body of margarine shall spread straight out of the fridge and leave the delicate honeycomb texture of fresh bread untorn.

Throughout history, ordinary people have known that a meal eaten in unison is an aesthetic experience, it binds together the family, as it also binds together the family of Man, breaking bread in amity. The colour, the texture, the aroma and flavour, the feeling of hard and soft, cold and hot, the music of crunching celery, or pork crackling, or game chips. Every sense is alive to the harmony. At the outset as he plods towards his goal, the food technologist, bowed under his load of scientific thinking, may not have realized what he was doing as he tinted ice-cream blocks three colours, standardized the viscosity of tomato ketchup and fitted together the varied geometry of mixed biscuits in those tins bearing on their lids the distant view of Windsor Castle in the twilight. The questioning 'consumerists' are harassed by doubts about the safety of the additives by which food beauty is to be attained and troubled by haunting fears that the very pursuit of elegance in this serious world may be a sin; yet they as much as the rest of us are being drawn, it seems to me, by the strange inexorable movement of what Bernard Shaw called the 'life force' to an understanding that science, one avenue towards the truth, made manifest through technology, itself impels us towards beauty. And if it is at first the beauty of a fish-finger or the delicate tint of a tub of soft margarine (a matter of deep concern to the manager of the soft margarine works), soon the cohorts of scientists, and the community who share the fish and the margarine with them, will alike come to know that all – or almost all – is in pursuit of beauty.

There are those who feel that science is a frigid force and that the technology by which it is applied to the industrial activities of the age dehumanizes communities that develop and use it. Deeper thought shows that this is not so. Consider one of the most sophisticated and esoteric of all industrial operations – the cold drawing of metal tubes. There are few communities even among the most advanced of the technological societies who can operate this process: the Swedes, to be sure, the Germans, the clever Japanese and the Americans too. Even within such advanced levels of industrialization, the units capable of carrying out the operation are few and those who work there are skilled. Admiringly, as a scientist in another field, I was taken through the great halls where the ends of big tubes are first battered to form a pointed handle; they are then grasped by great clamps, attached to huge chains and, held thus, are pulled by sheer force through orifices otherwise too small to let them pass. Under the tension applied, the metal stretches longer and thus the tubes, like elastic pipes, become narrower. After several passes through orifices of diminishing size, the stretched metal is annealed by heat to allow the atomic structure to relax.

Moving onwards through the works, I observed a different type of tube being subjected to a subtler and more complex stretching: my guides, revelling in their expertise, were obviously proud of it. Here indeed was advanced technology opening to them markets in new parts of the world. The tubes, it seemed, were aluminium, smelted from bauxite ore by the intense energy of an electric furnace, cold drawn by special machinery, shaped stepwise to attain a smaller diameter at one end than the other. What could be the purpose of such engineering skill to deserve so rich a return in international trade? To my naive inquiry came the reply: the tubes, cut to a length, threaded, polished and screwed together two by two were aluminium billiard cues. Grave Arabs in their white robes bending over the green cloth, intellectual Japanese, bustling Americans – all over the world were engaged in the coloured elegance of the game of snooker using cues of tubular aluminium. Perhaps there are other ends to industrial activity than anxious worry over an impending shortage of the world's natural resources. There are worse things to do in life than play billiards.

Life indeed is serious, and plans for an industrial future must surely be considered. Nevertheless the best-laid plans, even in the intellectual world of scientific forethought, may miss their mark as readily as did those of the rulers of the Roman Empire that lasted for a thousand years. Looking back on past ages we admire them most for the beauty they created, not for the useful things. Perhaps our successors will think little of the scientific achievement by which we developed detergents from petroleum, releasing for human food the great stocks of fat once needed to make soap. Instead they may admire, as I did on a visit to another great factory, the hairsprays and skin-cleansing creams, the shampoos and the toothpastes where the detergents that give them such efficacy as they possess, once made from petroleum, are now synthesized from edible fat. Most memorable of all was the department where bauxite, dug from distant places, was used not for smelting into billiard cues but to be compounded as aluminium salts into anti-perspirants. And the applicator, a container capped by a metal ball fabricated to precise tolerance to fit both the top of the vessel and the human armpit, was a testimony to precision engineering.

And the purpose of all was beauty and delight, as with great pictures and deathless music, the towering pinnacle of Salisbury Cathedral and the sonnets of Shakespeare. The toothpaste, the hairsprays, the deodorants, the shampoos, the very billiard cues themselves, into which so much skill and knowledge are being poured, are not strictly *necessary* for survival. Indeed, millions of our fellow creatures have lived and died unscathed without such things.

But let it not be thought that the pursuit of aesthetic excellence in coloured cake and sliced bread and in such curiosities as hairsprays and toothpastes are adequate targets for the scientifically oriented community of the future. It is apparent to those with eyes to see that the advancement of science and the continuing general adherence to it as a major avenue to truth and happiness, almost regardless of the fashionable unease of today, comes from a striving towards some higher excellence. Not long ago I attended a dinner of the Royal Astronomical Society. At table I sat next to a man of no apparently outstanding attainment: neither young nor old, he seemed little different from any other dinner companion. Yet this man of our own times, working in the laboratory of the Bell Telephone Company, had come specially from the New World to the Old to receive the Herschel Gold Medal: he had detected and measured the first shock of the mighty explosion by which – fifteen thousand million years ago, so the evidence implies – the universe began. The name of the man was A. Penzias. What greater goal can any society pursue than to further understanding of how and when the heavens and the earth came into being?

Index